AF871211

Praise for

GEOFF PERRY AND THE KETTERING GROUP

Des Welch has done a masterful job of capturing the unique talents and contributions of Geoff Perry and the Kettering Group that demystified Soviet space activities in the Cold War era.
—Marcia Smith, editor, spacepolicyonline.com

Geoff Perry was the quintessential brilliant English eccentric and a pioneer of what is now called 'open-source intelligence.' Welch takes us into the details of the detective work by which he and his army of schoolchildren transformed the Western public's understanding of the secretive Soviet space effort.
—Jonathan McDowell, author, *Jonathan's Space Report*

We value the pioneers in any field, but especially so in the sciences. Des Welch has researched widely to compile a detailed biography of this space pioneer and educator. It is long overdue.
—Stuart Eves, former member of the Kettering Group.

Des Welch presents a compelling narrative of a British physics teacher who set out to conduct contemporary, real-world demonstrations of fundamental principles at the dawn of the space-age.
—Mark Severance, NASA Human Space Flight Communications & Tracking Network Director (Ret.), and former member of The Kettering Group

Geoff Perry, c. 1985

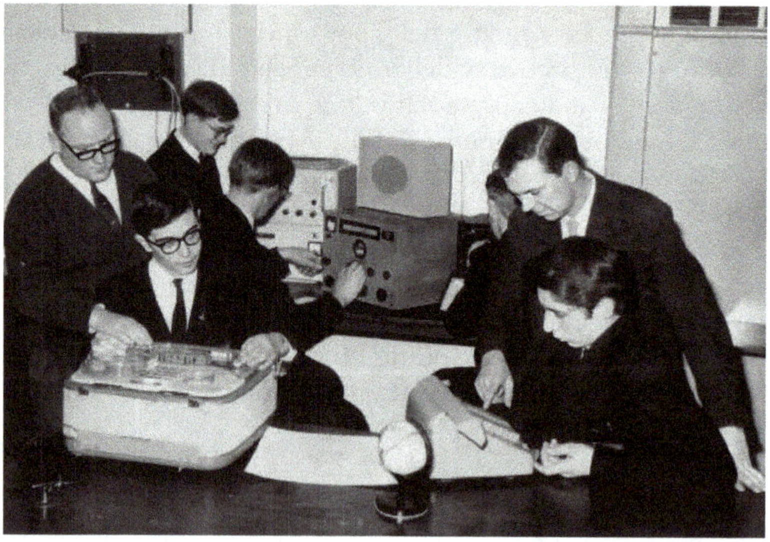

The Kettering Group at work in the Physics A Laboratory at the Kettering Grammar School, 1966. Left to right: Geoff Perry, Richard Jahn, Phillip Gregory, Robert Christy, Michael Sinnett, Derek Slater, Graham Coleman.

Geoff Perry and the Kettering Group

How a British teacher and his pupils exposed the Soviet Union and astonished the world

Desmond Welch

Europa Books • Sarasota

© 2026 by Desmond Welch

All rights reserved. No part of this book may be reproduced, distributed, or transmitted in any form or by any means without prior written permission of the author, except for brief quotations in reviews.

The right of Desmond Welch to be identified as the author of this work has been asserted in accordance with the Copyright, Designs and Patents Act 1988.

ISBN, eBook: 979-8-9966124-2-0
ISBN, paperback: 979-8-9966124-1-3
ISBN, hardcover: 979-8-9966124-0-6

Library of Congress Control Number: 2026944994

First edition

Appendixes B and C: Mark Severance
Front cover art: Tyler Jacobson
Graphics: Kate Welch
Cover and interior design: Karen Hood and Michael Vito Tosto
Editor: Hugh Barker

Europa Books, Sarasota, Florida
www.europabooks.press

To all the teachers around the world, present and past,

especially to my late mother and English teacher,

C. Patricia Welch.

Contents

Part III: Impact and Legacy

List of Abbreviations

AMSAT	Radio Amateur Satellite Corporation
ASTP	Apollo Soyuz Test Project
ATV	Associated Television
BASIC	Beginner's All-purpose Symbolic Instruction Code
BBC	British Broadcasting Company
BIS	British Interplanetary Society
BNCSR	British National Committee for Space Research
BST	British Summer Time
CIA	Central Intelligence Agency
COMSAT	Communications Satellite Corporation
CORONA	Program name for US photo-reconnaissance satellites
COSPAR	Committee on Space Research
CRS	Congressional Research Service
CWS	Cooperative Wholesale Society
DARA	Deutsche Agentur für Raumfahrtangelegenheiten
DLR	Deutsches Zentrum für Luft und Raumfahrt
DOS	Disk Operating System
DRA	Defence Research Agency
DSIR	Department of Scientific and Industrial Research
ELINT	Electronic intelligence
ESA	European Space Agency
FAI	Federation Aeronautique Internationale
FOBS	Fractional Orbital Bombardment System
FSW	Fanhui Shi Weixing, Recoverable Satellite
FTD	Foreign Technology Division
GLONASS	Global Navigation Satellite System
GMT	Greeenwich Mean Time
GPS	Global Positioning System
GRU	Glavnoye Razvedyvatelnoye Upravlenie
HF	High frequency
ICBM	Intercontinental Ballistic Missile
IGY	International Geophysical Year
IPY	International Polar Years
ISS	International Space Station
ITN	Independent Television News
ITV	Independent Television
KGB	Komitet Gosudarstvennoy Bezopasnost

KH	Key Hole
LOC	Library of Congress
MBE	Member of the Order of the British Empire
MHz	Megahertz, meaning one million cycles per second
NASA	National Aeronautics and Space Administration
NATO	North Atlantic Treaty Organization
Navsat	Navigation satellite
NAVSPASUR	Naval Space Surveillance System
NIE	National Intelligence Estimate
NORAD	North American Air Defense Command
NOSS	Naval Ocean Surveillance System
NOVA	US Public Broadcasting System science television series
OSCAR	Orbiting Satellite Carrying Amateur Radio
PC	Personal computer
RAE	Royal Aircraft Establishment
RAeC	Royal Aero Club
RAF	Royal Air Force
RAG	Raise and Give student charity campaign
RI	Royal Institution
RORSAT	Radar ocean surveillance satellite
RRS	Radio Research Station
RSRS	Radio and Space Research Station
S&T	Science and Technology
SIGINT	Signals intelligence
SIPRI	Stockholm International Peace Research Institute
STS	Space Transportation System
TASS	Telegraph Agency of the Soviet Union
UFO	Unidentified Flying Object
VHF	Very high frequency

Foreword

Inspired by the dawn of the space age and seeking to add excitement and relevance to his physics lessons at Kettering Grammar School, England, Geoff Perry made a tape recording of the signals received from Sputnik 4 in 1960 which demonstrated the Doppler effect. This was the catalyst for involving and engaging his pupils in satellite tracking. They received global attention when the 'Plesetsk' story broke in 1966. As time passed, like-minded adults from around the world contacted Geoff; several became Group members and over the years developed close friendships. Geoff came to be recognised as one of the world's foremost authorities on Soviet space activities.

From an early age, Geoff's daughter, Isabel, was an active participant in satellite activities, and her interest in the Kettering story continues to this day. Isabel, and her husband Charles, were first approached in 2023 by Des Welch, following an introduction from a Kettering Group member. As there was no existing publication concerning the history of the Kettering Group, Des believed it was a story worth telling. In his book, *Geoff Perry and The Kettering Group*, Des takes the reader through the Group's evolving story over four decades; he articulates their major achievements and announcements and includes anecdotes from those involved. All this content was a result of Des' detailed and comprehensive research, the majority from within the 'Geoffrey Perry Archive relating to the operation of the Kettering Group' held by the Science Museum, London, and interviews with Kettering Group members. The result is a coherent, fully illustrated compelling story which celebrates the achievement of Geoff Perry and The Kettering Group.

The revelations made by the Kettering Group and described in this book were enabled by Geoff's leadership and his eye for detail, supported through meticulous recording of information, application of scientific methodology and the co-operation between Group members and professional institutions. Geoff Perry was a remarkable man who led a group of enthusiasts, of which many were amateurs, in the pursuit of an understanding of the secrets of the Soviet Space Programme. This well-told narrative is a fine example that shows what can be achieved by a group of committed individuals and is highly recommended for anyone interested in following a similar approach or in learning about the history of the Soviet Space Programme.

—Isabel and Charles Carmichael

Preface

I first met a member of the Kettering Group in 1988 after I arrived at NASA's Johnson Space Center. Mark Severance worked as a Space Shuttle flight controller while I was a new instructor tasked with teaching the astronauts how to use the vehicle's controls and propulsion systems. I asked Mark a lot of questions, he patiently gave me answers, and we became close friends.

Occasionally, he mentioned something he was working on related to the Soviet Union's space activities, and that the leader of the group he was part of was a British secondary school physics teacher, Geoff Perry. Most of what he said went over my head as I was consumed by learning and teaching the Space Shuttle's complexities. One day, he told me the teacher had passed away and that he would be flying to England to attend Geoff's memorial service. Only then did I start to sense the importance of what Mark and his fellow Kettering Group members had been doing.

Over the intervening years, Mark described some of the revelations and deductions that Geoff and his network of collaborators and researchers had produced, and the spacecraft they had studied. When I toured an exhibit of Soviet cosmonautics in London in 2016, I realised much of the hardware on display was the subject of the Kettering Group's activities. I wanted to learn more about Geoff Perry and the Group, so I looked for a book about him. Finding none, I thought, 'There should be a book about this man and what he and his colleagues accomplished. I guess I need to write it,' and this is what I have done herein.

The more I learned about Geoff and the Kettering Group, the more excited I became about sharing his story. But this book is not a typical biography. It's more of a 'scientific biography' or a 'professional memoir', focused on the story of Geoff's teaching and satellite tracking activities throughout his career. It is also the story of many of the Group members, especially the ones I was able to interview. The Kettering Group was really two groups: a handful of pupils who worked on projects during the school year, and a network of domestic and overseas enthusiasts who helped Geoff gather and analyse data pertaining mostly to Soviet, and occasionally Chinese, space missions. The pupil portion of the Group existed from the early 1960s to 1984, when Geoff retired. The enthusiasts, mostly adults from many different walks of life, started working with Geoff in the mid-1960s and continued Kettering Group work until Geoff's death in 2000.

Partly because of secrecy, there was genuine mistrust on both sides of the Cold War. In elementary school in the early 1960s, I watched through our classroom window as a nuclear fallout shelter was installed in the front yard of a house across the street. While Geoff did not consider himself a Cold Warrior, he and his Kettering Group members shed significant light on the hidden purposes behind Soviet spacecraft. I'd like to think that Geoff and his colleagues helped make a terrifying situation a little less so.

NB: In the early years, the Group's official name was the Kettering Grammar School Satellite Tracking Group. However, in the book I often shorten the name to the Kettering Group (which was the official name dating from the mid-1970s on), or simply 'the Group.' Also, I am using the Russian language transliteration into English that Geoff used, especially the spelling of the word 'Cosmos' as opposed to 'Kosmos'.

Some detailed explanations of concepts or activities are set off in a separate box or as an appendix. These explanations can be skipped by the reader without impacting the narrative, but they will enhance understanding of how Geoff and his pupils accomplished their analyses.

This manuscript is based on published articles and books, interviews with participants, and publicly available material including archived documents and formerly secret or top-secret memos and reports that have been declassified by Western intelligence agencies.

Introduction

The Soviets and the CIA

The US-USSR superpower rivalry commanded the attention of the world in the 1950s and 1960s. Virtually any information about what the Soviets were doing militarily and in space was of keen interest in the West. Revealing Cold War secrets was not Geoff's motivation, at least not initially. But he did love a puzzle, and that, as much as anything, drove his curiosity and analysis of the goings-on behind the Iron Curtain.

Geoff Perry was a grammar school science master (corresponding to a US high school physics teacher, an English grammar school being the equivalent of a high school in the US). Geoff's initial desire was to use a recording of a satellite's transmission to illustrate a physical principle, the Doppler effect, for his pupils. That desire developed into a lifelong passion to study satellites. While some tracking occurred during the school day, most of everything described in this book he did outside of his teaching role. He was an inspiration and mentor to many of his pupils. Geoff's band of enthusiasts around the world still feel his influence on their lives today.

Much of the Kettering Group's activities took place during the height of the Cold War between the then-Soviet Union and the NATO powers, primarily the United States. For those unfamiliar with the intensity of the Cold War, I describe how the US-Soviet rivalry formed the backdrop to the Kettering Group's activities. From there, I describe what Geoff was discovering about the Soviet space programme related to his investigation of satellites launched by the Soviet Union, especially those from the Plesetsk launch site; what the Soviets were actually doing at Plesetsk; and what Western intelligence, primarily the CIA, knew about the existence and function of the secret site.

But Geoff's career didn't end at the reveal of Plesetsk. Continuing to track, monitor and analyse Soviet space operations led him to become a world-renowned expert on what the Soviets were up to in orbit. Geoff was regularly published and quoted in major newspapers and journals around the world. I've collated his activities and accomplishments into chapters on his analyses of Soviet satellites, his television and radio appearances, and his many other high-visibility activities. The reader will notice an emphasis on Geoff's articles and journal papers. He firmly believed that publication of one's findings was an important part of the scientific method, and he set a clear example of that for his pupils.

Why did Geoff Perry—a British secondary school physics teacher whose primary motivation was providing real-world demonstrations of basic principles of physics such as the Doppler shift—become so engrossed by what the Soviets were doing in space? And why were his revelations greeted with such interest and publicity? To help answer these questions, we must look briefly at the history of space activities behind the Iron Curtain.

The Soviet Union's rocketry programme made rapid advances in the 1940s and 1950s, aided in part by a contingent of German scientists and technicians relocated to Russia after World War II. In 1953, Soviet planners ordered the development of a missile that would carry a 5,500 kilogram (12,000 pound) thermonuclear bomb 8,000 kilometres (5,000 miles)—the world's first intercontinental ballistic missile (ICBM). Since the Soviets' mammoth nuclear warheads couldn't be reduced in mass at the time, a powerful rocket was needed.[1]

Design and testing of the Soviets' new ICBM lasted two years. Several launch attempts in 1957 from Baikonur, their main launch facility in Central Asia (although the village of Tyuratam was closer to the actual launch site, so Western sources use that name), ended in failure. But in August, the rocket—designated the R-7 or informally, Semyorka, 'Old Number 7'—flew successfully to its planned target, though the missile's simulated warhead disintegrated late in the entry phase.[2]

When the Soviets announced their satellite plans for the International Geophysical Year (IGY, see below), Western experts were not impressed. The Soviet Union's technological abilities were considered backwards and unsophisticated by US and European standards—erroneously, it turned out. Sergei Korolev, the Soviet Union's space programme chief designer, had been pushing for an Earth-orbiting satellite launch using the R-7 booster. Initially, the Soviet Council of Ministers resisted the idea, but Korolev finally got Premier Nikita Khrushchev's blessing, on the condition that progress on ICBM development wasn't affected.[3]

IGY and Sputnik

In the early 1950s, US scientists were advocating for an international scientific research project to study the Earth's surface, its atmosphere and the space environment beyond. This wide-ranging effort would build on the previous International Polar Years (IPY), conducted in 1882–83 and in 1932–33. The first IPY involved several European countries, along with the United States and Russia, coordinating efforts to study both the Arctic and Antarctic environments. The second IPY included more than forty countries and expanded research to include radio phenomena and upper atmosphere properties. With a wider focus than the earlier studies, the new effort was dubbed the International Geophysical Year, to run from mid-1957 to the end of 1958.

In 1955, the Soviet Union announced a plan to send an artificial satellite into Earth orbit as part of IGY; the United States publicised a similar plan shortly afterwards. The world's press trumpeted the US announcement, while the Soviet Union's proposal was virtually ignored—the Western world being convinced of its technological superiority. The Soviets intended to launch a large, instrument-laden spacecraft to support its IGY goals, but the satellite had numerous birthing pains, so they substituted a smaller, simpler craft: Sputnik.[4]

After discarding a more complex design, Korolev and his team produced the 'Simplest Satellite 1', or, in Russian, 'Prostreishiy Sputnik 1.'[5] This name was mostly unknown in the West, but when the Soviet TASS news agency's announcement of the launch called the spacecraft 'the first artificial Earth satellite', Western media latched on to the Russian word for satellite, 'sputnik' (literally, 'fellow traveller'). The 82-kilogram (180-pound) sphere contained a couple of radio transmitters, a battery, and little else; Sputnik transmissions were in the high frequency (HF) radio band that nearly any amateur radio operator in the world could receive.

The Soviets were focused on beating the United States into space, which they did when Sputnik was launched on 4 October 1957—the first major event in what became known as the 'space race' between the United States and the Soviet Union. But while USSR authorities did not appreciate the true impact of their feat for many weeks, Western powers and their citizens grew increasingly concerned about the potential for Soviet nuclear bombs raining down from the sky.

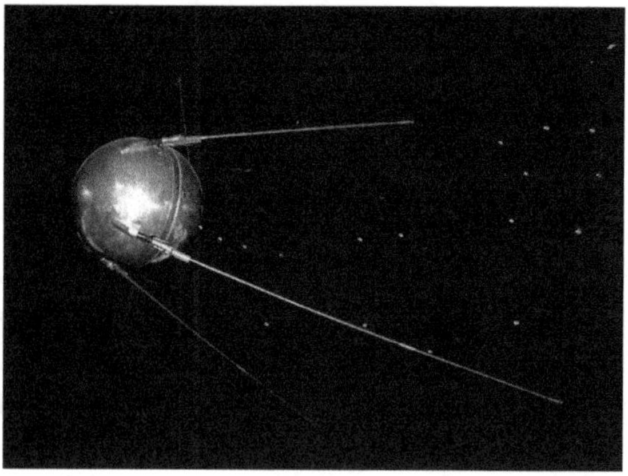

Sputnik model, the Museum of Space and Missile Technology, Saint Petersburg, Russia.

The same year they announced the intention to launch Sputnik, Soviet planners decided to build a new ICBM test and launch site in northwest Russia, south of the port city of Archangelsk, near a small village called Plesetsk.[6]

The launch of Sputnik may have astonished US citizens and many others around the world, but it did not surprise the Western intelligence community. In late 1955, the United States' Central Intelligence Agency (CIA) had published a classified report warning that the Soviets could place in orbit 'a relatively uninstrumented [sic] vehicle by 1958'.[7] Using radar, radio listening posts and other means of information collection, CIA intelligence on Soviet rocketry was continually being updated throughout the mid-1950s. Predictions about the Soviets' capability to send a satellite into orbit were refined and regularly presented to US President Dwight Eisenhower. At one point, the CIA assessed that the Soviets could launch their first satellite as soon as early 1957.[8]

The US intelligence community was also concerned about the propaganda triumph for the USSR if they succeeded in launching the world's first orbital craft. A report to the CIA director warned of a 'psychological defeat were the Soviets to launch a satellite before the United States.' Despite the CIA's warnings, most US leaders did not believe the Soviets had the technology to beat the US into space. The country's leaders and much of the US population thought that 'Russians are thick-fingered peasants, with terrible electronics capabilities, and that they are actually fumbling their way about in space,' as author Martin Caidin put it.[9] No less than US Defense Secretary Charles Wilson dismissed the Sputnik launch as a 'scientific trick.'[10]

President Eisenhower said nothing for several days after Sputnik reached orbit, while newspaper headlines around the country reflected citizens' growing worries about Soviet capabilities and intentions. Five days after Sputnik's bleeps were first heard, the President famously said the launch did not increase his apprehension for the security of the nation 'by one iota.'[11] The US public, who were aware that the Soviets had both atomic and hydrogen bombs, increasingly failed to share Eisenhower's optimistic outlook. Eventually, US leaders acknowledged the Soviets' technological prowess that only weeks earlier they had scoffed at.

CIA reports describing the Soviet Union's progress in rocketry and satellites continued to flow across the President's desk over the next two years. A September 1959 National Intelligence Estimate (NIE) report assessed that the Soviets almost certainly had the capability to launch nuclear weapons via ICBMs that could reach the US. It also specified that the only known Soviet missile launch sites were Baikonur, in the Kazakhstan desert, and Kapustin Yar, about 1,000 kilometres (620 miles) southeast of Moscow.[12] The USSR had launched its first ballistic missile from Kapustin Yar in 1947. While the Soviets had publicly revealed the Baikonur site, they considered the existence of Kapustin Yar secret. However, Western intelligence and media had long since identified it.[13]

As early as 1957, the CIA suspected that other ICBM sites would be developed. In a classified report, CIA analysts suggested one likely area for a missile facility was the Kola Peninsula in northwest Russia. The region was closer to the eastern US than other launch sites, allowing for shorter, more accurate ICBM flight trajectories. But the area was near Finland and Norway, which could—and later did—enable Western monitoring of Soviet activities there.[14]

The US' National Security Agency published intelligence reports in 1958 indicating possible military construction activity in northwest Russia. By 1959, intelligence collection confirmed that an ICBM launch facility was being built in the Kola Peninsula region, having started as much as a year earlier. The CIA targeted several sites in the USSR for photography from a U-2 spy aircraft on 1 May 1960, despite warnings of improved Soviet surface-to-air defences. One of the most important sites for the mission was the area around Plesetsk, a small village some 200 kilometres (125 miles) south of the White Sea, where signals picked up by a Norwegian listening post had indicated significant military-related activity.[15]

The pilot of the U-2 on the 1 May mission was Francis Gary Powers, but he never made it to Plesetsk. He was shot down by a Soviet air defence missile while cruising at approximately 21,000 metres (70,000 feet) over the heart of the USSR. Powers bailed out of his airplane successfully and was arrested. The incident was a national embarrassment for the United States; it also significantly disrupted US-Soviet relations.

In August 1960, Powers was sentenced to ten years in prison after a Soviet show trial. (He was released in 1962 during a prisoner swap with the US.) On the front page of the *New York Times* the following day, next to the Powers trial story, the US Air Force announced the first-ever mid-air retrieval of a capsule ejected from Discoverer XIV, allegedly a civilian science satellite.[16] But rather than returning scientific data, the capsule contained thousands of feet of film depicting strategic military targets throughout the Soviet Union, bounty from the first successful—and secret—CORONA reconnaissance satellite. In one mission CORONA had provided more photographic coverage of the Soviet Union than all previous U-2 overflights combined.[17] The Soviets would soon start orbiting their own spy satellites, which Geoff and his pupils would avidly study.

The Soviets had managed to keep their R-7 rocket, which lifted Sputnik into space, secret for several years. An August 1957 photograph, though, from a U-2 did show a clear image of R-7 facilities at Baikonur, with an R-7 rocket being readied for launch; however, its capabilities apparently could not be assessed from the image.[18]

At the 1967 Paris Airshow, the Soviets decided to show off their primary satellite booster and what had been, for a short period, their main ICBM platform for nuclear weapon payloads. Western observers were surprised by the large size and rugged nature of the R-7 booster. Although the West had caught up to the Soviets with its own fleet of ICBM vehicles by then, experts realised that Soviet engineering had been forced to develop high thrust rockets to accommodate their massive atomic weapons. In the mid to late 1950s, the Soviets had not yet mastered miniaturised electronics to the degree that the US had.

Why were the Soviets so much more secretive about their spaceflight activities than the United States? Granted, the US covered up some of its satellites with misleading names, but the Soviets tended to divulge almost nothing about any of their space missions until they were in orbit and appeared to be successfully carrying out their programming, classified or not. Even then, the Soviets described most missions in the vaguest of terms. So, as will be seen in later chapters, any penetration of the Soviet censorship was a big deal in the West.

Secrecy in the Soviet Union was already commonplace at the dawn of the Space Age. The term 'Iron Curtain,' popularised by Winston Churchill in a 1946 speech, referred to the blockade of information into or out of the USSR, as well as to the virtual barring of citizens from leaving communist countries. The reasons for Soviet secrecy were partly similar to those of the United States and other countries: 'The defensive motive of screening vulnerabilities against their exploitation from abroad, and the offensive motive of furthering national objectives through deception and surprise,' according to a US government study.[19]

The roots of secrecy, though, extend far back in Russian history. Russia had been a crossroads for marauding forces for centuries. As a result, at the heart of Russian culture, there was a deep distrust of foreigners.[20] The numerous defeats by various invading armies generated a sense of insecurity and a feeling of backwardness—citizens sensed the lack of technology and know-how that was needed to keep the country safe.

The rise of communism exploited the Russian mentality and added yet more reasons for secrecy: the communist tendency to suspect and foment conspiracy; the desire to prevent Soviet citizens from comparing their standard of living to that of the Western world; insistence on the appearance of continual progress and success, no matter what the reality of events was; and internal censorship intent on preventing any criticism of leadership or dogma.[21] Ultimately the Soviets wanted to keep secret any progress—or lack of progress—around their space efforts, until they had a positive result to announce to the world. They also preferred to keep their goals for the space programme confidential, to prevent their rivals from accomplishing something before they did.

The USSR, of course, wasn't the only country that kept secrets about its space programme. In early 1985, the *Washington Post* published an extensive analysis of secrecy around US military space activities. The authors argued that criticism of an earlier *Post* article describing the secret payload of an upcoming Space Shuttle mission was unjustified. They pointed out that unclassified sources such as Congressional hearings, scientific conferences and journals such as *Spaceflight* and *Aviation Week and Space Technology* routinely disclose information about both US and Soviet military efforts in space.

The authors detailed the coverage of previous classified military launches including electronic and communications intelligence systems, and how that information had been derived. They chided the US government's reticence to reveal basic details about a military satellite, such as launch time. The authors noted:

> Each superpower's space monitoring networks include detection and tracking radars, telescopes that can see out to geosynchronous altitudes, ships equipped with electronic eavesdropping equipment, satellites and land-based antenna 'farms' that intercept data transmitted between satellites and ground stations. Even private individuals armed with basic radio gear have managed to decipher large portions of Soviet space operations. For years, the recently retired Geoffrey Perry, with the aid of his students at the Kettering Grammar School in England, [has] monitored the transmissions to and from Soviet satellites. With the aid of this data Perry has written numerous authoritative articles on the Soviet military space programme, including Soviet photographic reconnaissance and signals intelligence satellites. It seems likely, even inevitable, that the Soviets can do at least as well deciphering the U.S. programme as those schoolboys, particularly when operating against a far more open society.[22]

The article concluded, 'This entire Shuttle incident indicates a disturbing respect for Soviet secrecy practices, and an unfortunate desire to mimic them.'[23]

Modern day western Russia.

Part I

Origins and Plesetsk

1

Secret Base, Secret Satellite

1966

By 1966, Geoff Perry and his pupils had been listening to satellites for several years and knew what a typical Soviet spacecraft signal sounded like. But there was something different about the latest transmission they had picked up. Although the Group did not know it yet, they were starting down the path to their greatest achievement and to world-wide fame.

The Soviet Union launched a satellite in March 1966 from a clandestine missile base in northwest Russia. Called Cosmos 112, it was the latest in a series of Soviet spacecraft that were purportedly scientific in nature. But the 'Cosmos' label was intentionally misleading; the payload was a military reconnaissance satellite.[24] The Soviet Union had sent many photographic spy missions to space from its busiest launch facility, Baikonur, located east of the Aral Sea in what was then the Kazakh Soviet Socialist Republic. Cosmos 112 was the first orbital launch from Plesetsk, an intercontinental ballistic missile site that was ready to fire nuclear weapons towards United States targets in wartime.[25] When it lifted off, the Soviets had not announced the Cosmos 112 launch, nor what its mission was, nor the fact it was being launched from a classified facility near the Arctic Circle.

In the mid-1950s, the looming threat of Soviet nuclear-tipped intercontinental ballistic missiles had prompted the United States intelligence community to deploy powerful radars, electronic eavesdropping equipment and the means to process and analyse reams of data from multiple sensors around the world. US allies—Norway and Turkey in particular—had provided key locations, enabling the reception of signals revealing Soviet missile and satellite activity.[26] The Norwegian listening post near the Russian border at Korpfjell, about 850 kilometres (530 miles) northwest of Plesetsk, would have been the nearest Western listening post to hear the high frequency telemetry signals from the Cosmos 112 spacecraft during its ascent into orbit.[27]

Telemetry

The purpose of most artificial satellites is to gather information, whether for scientific, commercial or military purposes, and relay that information to a ground station where it can be processed and distributed to users. Since there is no physical connection to an orbiting satellite, the information produced by the satellite's sensors must be sent by wireless methods.

Most commonly, a radio transmitter on board the satellite sends a signal to a receiving station on the ground. The signal is encoded with sensor data, which is decoded after reception. This system of sending and receiving data from a remote source is called telemetry, from the Greek words 'tele' (a long distance away) and 'metria' (the process of measurement).

A continent away, in the Aleutian Island chain southwest of the Alaskan peninsula, Western intelligence personnel at Shemya Air Force Station operated three large radar antennas that looked for anything flying above the atmosphere. The radars and their associated electronics could find, process and help catalogue objects in space that were as small as one metre (three feet) across.[28] Cosmos 112 was detected as soon as it rose above the horizon in its orbit. Information about its flight path was stored for later transmission. Other radars in the US's worldwide space surveillance network were undoubtedly alerted. Cosmos 112 made two more passes over Shemya before the following morning, and Shemya's radar provided additional data through the night to help refine the calculation of the satellite's orbital path.[29]

Just 90 minutes after the launch of Cosmos 112, in a physics laboratory at Kettering Grammar School, pupils had powered up a shortwave receiver and related equipment as they did every school lunch break. They were members of the Kettering Grammar School Satellite Tracking Group. Leading the Group was Geoff Perry, Head of Physics at the school. Membership in the Group was selective; you were a member if Geoff invited you to join and you accepted. Former first form pupil (first year of middle school) David Dean described how he became a member.

> I approached Mr. Perry and said I would like to join the satellite club….He looked at me and said, 'Dean, this is not a club that anyone can join, you're asked to join, by invitation only.' So that put me in my place. What I found out later is he was looking to see who was doing well in physics. So it took the whole of my first year at the grammar school for me to prove myself. At the end of the first year, he asked me whether I'd be interested in joining and I of course leapt at the chance.[30]

The only non-Kettering Grammar School member of the Group in 1966 was a Swedish university student, Sven Grahn, whom we shall meet later.

The original purpose of the Kettering Group was to support the collection and analysis of data gathered from satellite transmissions, which reinforced methods associated with the best practices of scientific research. Geoff and the Group were driven by the desire to discover the purpose of Soviet spacecraft orbiting the Earth about which little or nothing had been revealed. The equipment the Group employed to tease out the Soviet spacecraft transmissions was tucked away in a corner of Geoff's physics laboratory.

The Soviets' use of the high frequency radio band (3–30 megahertz, or MHz) enabled Geoff and his pupils to easily monitor satellite telemetry; their location in England was advantageous for listening to the USSR's satellites as they passed over the UK and came in range of ground stations inside the Soviet Union. Occasionally, with the right propagation conditions, they could hear Soviet spacecraft during their ascent phase, once their rocket's aerodynamic fairing was jettisoned several minutes after launch and the spacecraft's antennas were deployed. While Cosmos 112 flew just a few hundred miles southeast of Kettering on its first orbit, though, the Group members did not hear it transmitting, perhaps because the satellite's transmitter had not yet been activated by its control centre.[31]

Over the years, Geoff and the Group had assembled an array of equipment that could receive and process information from the Soviet Union's spacecraft, including the Cosmos series of satellites. But their setup was nothing like the sophisticated equipment at Western space surveillance facilities. Consisting of a vintage high frequency radio receiver and a few other pieces of hardware, their electronics collection was simple yet effective.

The frequencies that Cosmos satellites transmitted on, mostly around 20 MHz, are also known as shortwave signals. US satellites typically transmitted on frequencies that were outside the reception capability of the old Marconi CR100 shortwave receiver that Geoff's col--eague and Kettering Group member, Derek Slater—the school's chemistry teacher—had loaned to the Group. So the pupils were usually limited to monitoring Soviet spacecraft. Derek had borrowed a signal generator to use as a stable oscillator; the signals received from the satellite and the local oscillator combined to produce an audible tone in a loudspeaker.

The equipment also included a reel-to-reel tape recorder for preserving satellite transmissions for later analysis, a mechanical calculator to work out orbits and ground tracks, and a toy globe Geoff had borrowed from his daughter Isabel, fitted with a copper loop to indicate a spacecraft's orbit around the Earth.[32] The radio's antenna was a simple dipole, essentially a piece of wire strung between two school buildings using cable that normally connected a television set to its antenna. Geoff's intellect, attention to detail and persistence, along with the technical expertise of Derek, were the Kettering Group's secret ingredients.

Isabel's toy globe with a copper loop attached to depict a satellite's orbit.

Marconi

The heart of the Kettering Group's tracking equipment in 1966 was the Marconi CR100 receiver. Guglielmo Marconi, considered the inventor of radio, founded the Wireless Telegraph and Signal Company in 1897, later known as British Marconi. Before and during World War II, Marconi's company produced the CR100 shortwave receiver, which saw service in all military branches of the UK. The CR100 was rugged and relatively easy to use; many units were employed by British operators during the war relaying encrypted Nazi communications to Bletchley Park, the British facility that cracked enemy codes. The receiver was also used by the UK's Special Operations Executive listening stations for wartime communication with Allied intelligence agents in Europe.

The CR100 was widely available on the UK surplus military equipment market in the 1950s. Derek Slater's Marconi receiver was manufactured in 1937; he purchased it in 1958 for £25 ($70).[33]

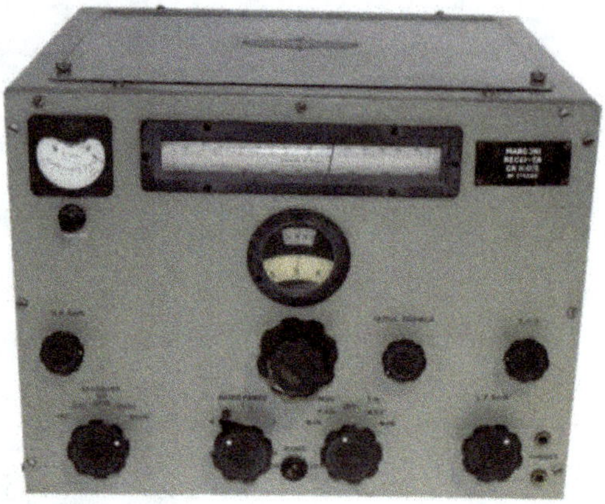

Pre-WWII Marconi CR100 shortwave receiver used for reception of satellite signals at Kettering Grammar School.

During its second revolution, Cosmos 112 spacecraft passed 950 kilometres (650 miles) northwest of Kettering, soared over Scandinavia, crossed the Barents Sea, and arced southeast across the central Siberian plateau.[34] From the satellite's control centre near Plesetsk, a radio signal to the spacecraft would have turned on its transmitter as it crossed Russia.

At this point, Cosmos 112 was 6,500 kilometres (4,000 miles) east of Kettering and well beyond line-of-sight to the British Isles. But a funny thing can happen to shortwave signals: the radio waves may bounce back and forth between the ground and the ionosphere—a layer of the atmosphere from 80 to 970 kilometres (50 to 600 miles) above the Earth, where the Sun's ultraviolet radiation strips electrons from atmospheric molecules, creating electrically charged particles. In the right conditions, the ionosphere can act as a mirror for shortwave radio transmission, allowing the signal to travel much farther than line-of-sight distance. This phenomenon allows communication across many thousands of miles, sometimes around the world.[35]

From central Asia, the first faint transmission from Cosmos 112 was heard on the Kettering Group's receiver between 1346 and 1348:30 Greenwich Mean Time (GMT). However, the pupils did not yet know what spacecraft they were listening to nor its significance.

2

Geoff, A Portrait

Geoff's love of teaching and learning helped put him, as they say, in the right place at the right time. For several decades, no one else in the world—who wasn't locked away behind classified doors—could do what he did in revealing the hidden activities and purposes of the Soviet space programme. People felt that Geoff was one of those big, bright personalities who seemingly mastered everything.

Geoffrey Edward Perry's first encounter with a missile was not his most pleasant, but the event foreshadowed a life-long fascination with rockets and satellites. One day in 1944, as World War II was raging across the European continent, 17-year-old Geoff heard a couple of loud bangs just before a large explosion occurred a few miles from his home in southeast England; the blast was so strong, several items on a shelf in his room were knocked to the floor and shattered. When it was later determined that the explosion had been caused by a Nazi V-2 missile (the two bangs were a double sonic boom from the missile), he 'realised that space travel was no longer a wild dream.'[36]

Born in 1927, Geoff grew up in the bucolic surroundings of Braintree, Essex, during the depression of the 1930s. His father, George Edward Perry, served in the Royal Artillery and had an astute mathematical mind. His mother Rose worked in a local shop, had a keen intellect, and enjoyed needlework, letter writing and singing. Geoff was an excellent pupil at Braintree County High School; in 1945 he achieved his Higher School certificate, won a Special Subject Prize in Mathematics and received an Essex County Major Scholarship in Science at the University of Reading.

Geoff, left, with his parents, Rose and George, c. 1949.

Turning 18 just after the end of World War II, he was not subject to military conscription due to the British armed forces post-war draw-down. Whilst at school, though, Geoff participated in the Army Cadet Force, a youth organization focused on adventure, leadership and outdoor skills. The Army Cadet Force, sponsored by the Ministry of Defence, supported the UK's Home Guard during the war. He graduated from the University of Reading having read physics and mathematics.

Geoff (second from right) with fellow Army Cadet Force members, c. 1945.

Geoff was passionate about precise record keeping from a young age. During the summer after the war ended, he organised a bike ride around southern England with a friend. Demonstrating an eye for detail and his data collection methods, he kept track of all activities and expenses, logging everything in a small notebook. 'Daddy was always observing and making notes,' his daughter Isabel remembered. 'He wrote with perfect penmanship and with no spelling errors. It was just in him.'[37] He wrote everything down—events, phone calls, arrival and departure times, anything of interest that occurred in his life. A Kettering Group member recalled later, 'I learned from Geoff [that] if something happens, write it down. It might not seem relevant, but five years later you might go, ah, I remember when this happened before.'[38]

Of his career ambitions, Geoff later said, 'I always wanted to be a schoolteacher; my cousin is a schoolteacher and my aunt is a schoolteacher.…So when I went to university there was no shadow of a doubt I was going to be a physics teacher.'[39] Indeed, after graduating from university, he taught at a private school in Gloucestershire, before returning to the University of Reading for his Postgraduate Certificate of Education. Following this, he took up the post of assistant master for chemistry, physics and mathematics at his former high school in Braintree before moving to Kettering in 1954. Geoff met Jean Brand through the Baptist Church in Braintree where they shared a love of music; Jean played the organ and Geoff sang and was a clarinettist. He proposed to Jean in 1953, and they married the following year. They had a daughter, Isabel, in 1956, who followed in her father's footsteps to become Head of Science and Physics at an independent girls' school in Malvern, Worcestershire.

While teaching at Braintree County High School, Geoff received a gift that would help set him on his path. For Christmas 1951, a cousin gave Geoff *The Exploration of Space* by Arthur C. Clarke. Geoff later wrote that the book was a milestone in prompting his interest in rockets and satellites, especially since it 'set out the principles [of spaceflight] based on fundamental physics.'[40]

Not only did Geoff play the clarinet, but he was also an accomplished trumpet player.

Arthur C. Clarke

In the 1950s, Arthur C. Clarke was a noted UK authority on spaceflight. He had previously written *Interplanetary Flight*, the first book in English to describe astronautics in detail to a general audience. Clarke is also credited with popularising the idea of geostationary communications satellites in a 1945 *Wireless World* magazine article.[41] Along with publishing many non-fiction books, he would later enjoy fame as a science fiction author. He was author of the novel *2001: A Space Odyssey* and co-writer of the screenplay for the film.

Clarke had obtained a degree in physics and mathematics at university. Both he and Geoff were involved in the British Interplanetary Society (BIS), Clarke being chairman in the late 1940s and early 1950s. The two did not know each other well, though they met in the 1960s—Isabel has Clarke's signature in her autograph book—and again in the 1980s. But both men, of British heritage and similar courses of study, provided their own unique influences on spaceflight throughout their lives.

In 1954, Geoff accepted a science teaching position at Kettering Grammar School in the county of Northamptonshire. The town of Kettering, an hour north of London by train, was a bustling manufacturing centre in the 1950s, known for producing boots and shoes; the town's origins can be traced back to Roman Britain times. Queen Elizabeth I had founded the Kettering Grammar School in 1577; by the mid-20th century, it was a selective school for university-bound boys. The nearby Kettering High School was the all-female-pupil equivalent.

Geoff's interest in rocketry was further spurred on by the upcoming International Geophysical Year. Both the US and the Soviet Union intended to launch the first artificial Earth satellites as part of their IGY research efforts. Thinking that his pupils might be questioned on these projects in their university entrance exams, Geoff carefully studied the latest plans for the potential spacecraft and made sure his pupils were prepared. Because of his interest in the subject, Geoff declared later that he was 'not in the least surprised when the Soviets launched Sputnik' in 1957.[42]

Throughout his life, Geoff had numerous interests and was always engaged in one activity or another. Isabel recalled how much her father liked to travel, organizing regular trips that the family called Perry Tours. 'We visited many European cities for a short break during my half terms. We went everywhere. He would organise it and knew exactly what needed to be seen. It was absolutely marvellous.'[43] Charles Carmichael, Isabel's husband, added, 'Not only did he work out what should be seen, but he ran a programme with no gaps in it whatsoever.'[44] With a laugh, Isabel added, 'We returned home absolutely exhausted.'[45] Geoff also enjoyed hitching up the family's caravan (travel trailer) and touring around the English countryside during school breaks. In the long summer holidays, the Perry family would park their caravan on the Cornwall coast for several weeks.

Geoff collected stamps throughout most of his life. His collection, which included many pages of space-themed stamps, fit in with his love of cataloguing and recording most everything in his life. Geoff recalled later, 'The Russian stamps are very accurate in the way they're drawn. We learned from [a stamp depicting the] Soyuz 9 [crewed space mission] in fact that there was no rendezvous radar. It wasn't there for Soyuz 9 because it had nothing to rendezvous with.'[46]

Part of Geoff's space stamp collection.

Geoff was also an avid bird watcher. Kettering Group member Mark Severance once asked him why he enjoyed the hobby: 'He said there was somebody at Kettering who said you can tell from migratory patterns what the weather is going to be like, and it's a great tipoff from nature that certain things are about to happen. He thought, 'That sounds kind of important, I guess I should start doing that.'[47]

Geoff birdwatching, 1986.

Sport was another passion of Geoff's. He was very competitive and in school enjoyed cricket, running cross-country and swimming. As an adult, he followed football and cricket and was a 'accomplished spin bowler', according to Isabel. Family friend and Kettering Group member Sarah Mobbs remembered, 'There was a squash court in Bude [where the Perrys vacationed in summer], and he said, "I'll give you a game of squash." He beat me hands down! He said, "I taught you about angles and things. You're not playing like a scientist; you're just whacking it." He worked out every single shot, where the ball would go. But that was Geoff. So, we had another go and I translated what he said, but he still beat me.'[48]

Geoff was also known for his sharp sense of humour. Lectures by Geoff were often punctuated by bursts of laughter from the audience. In a letter to colleague Desmond King-Hele regarding Soviet eight-day mission operations, Geoff joked, 'I feel these Saturday launches are part of a diabolical two-part plot by the USSR. They know that if they launch on Saturday, [King-Hele] will not be told and the subsequent Sunday recovery means that [I] lose my lie-in!'[49]

Isabel recalled a letter Geoff had received with a cancellation mark that looked like eyeglasses across the face on the stamp, the face being one of Henry VIII's wives. Geoff wrote on the envelope, 'I wasn't aware that Anne of Cleves had defective vision.'[50]

Sarah Mobbs recalled some of Geoff's other interests. 'He was always interested in so many things. Rock pooling in Bude was always fun, he'd take you along and show you all sorts of things like molluscs. We used to do body surfing. He loved that. My dad and I used to laugh that Geoff would always know which wave was the best one. 'That one there!' He would go around the beach and collect as only Geoff could. You know the lolly [popsicle] sticks? He liked them so he could label the sowing of his seeds back in his garden. He used to write on them. He wouldn't let anything go to waste at all.'[51]

Geoff was also famous for scooping up freebies at business shows. Charles Carmichael recalled, 'We went to a lunch down in Cornwall for some business which I did from time to time. On this particular occasion there was a business exhibition in the hotel. We went into that to have a look round. He came out laden with stuff. Notepads, pens, pencils, and all kinds. And then Jean could go round and get the same things. Geoff would tell her, "Go there, there and there, the best places."'

Isabel often says about her father's extensive activities, 'I don't know how he did it all!' Along with his boundless energy, Geoff's biggest asset was his wife, Jean. Jean was his rock, his 'support with a capital S,' recalled Isabel. 'Logging all the phone calls, driving him to lectures, helping him with paperwork and checking data to help him to update *The RAE* [Royal Aircraft Establishment] *Table of Earth Satellites* [book]. So, they would have sat at the kitchen table reading figures backwards and forwards to each other, because it is only too easy not to spot your own mistakes, isn't it? Absolutely, he couldn't have achieved his success without her.' Charles added, 'It became a way of life, and Jean was integral and central to that way of life.'[52]

Jean and Geoff, 1993

3

A New Sound

1957 – 1961

Geoff was nothing if not persistent. Here, he recognised an opportunity but couldn't bring it to fruition himself. However, when circumstances changed, he saw a way to achieve his objective. And he formed a lifelong friendship in the process.

In his science master position at Kettering Grammar School, Geoff taught basic physics across the whole pupil age range. At first, he relied on tried and tested methods of describing motion, forces, waves, and other principles of physics. But he was soon searching for more novel ways to stimulate his pupils' interest in physics lessons. While attending the Annual Meeting of the Science Masters' Association in Leeds in December 1957, Geoff attended a presentation that changed the course of his career. A fellow science master and ham radio operator from Stowe School, John Osborne, played a tape recording of the HF transmissions from Sputnik, which featured a dramatic change in the pitch of the spacecraft's signal.[53]

A radio signal from a satellite has a higher frequency as the satellite approaches a receiver and a lower frequency as the satellite moves away in its orbit. This is analogous to the sound heard as an ambulance siren moves toward and travels past a listener. The change in frequency due to relative motion is known the Doppler effect. When Geoff heard Osborne's recording, he noted that the whole-octave Doppler shift in pitch was remarkable. He realised the impact it would have if he demonstrated the Doppler effect to his pupils with a sound ripped from the headlines, as the bleeping of Sputnik, the first artificial satellite, had gripped the world.

After the talk, Geoff asked Osborne if he could have a duplicate of the Sputnik transmission recording. Osborne said he would send a copy, but he never did. For Geoff, this turned out to be a most fortuitous unfulfilled promise; if the tape had arrived, the rest of Geoff's story may not have occurred.

By the summer of 1958, Geoff had still not received John Osborne's promised recording. He didn't have access to the equipment to make his own recording, but a solution presented itself that August when Derek Slater, a chemistry teacher and radio amateur, joined Kettering Grammar School. Together they planned to record signals from the next Sputnik to be launched.

With an anticipated pay rise in his new job, Derek had purchased the Marconi shortwave receiver. He was intent on establishing a ham radio club at Kettering Grammar School and building an antenna on campus for the receiver. Moreover, Derek had experience in the Royal Air Force with shortwave equipment, having frequently monitored during his service what he called 'the mush and crackle' of a distant high frequency signal.[54]

Shortwave Radio

Radio communication was pioneered by Guglielmo Marconi and others at the beginning of the 1900s. Improvements to amplification and tuning circuitry enabled experimenters to build their own equipment and contact each other, first via Morse code and later by voice.

In the early 20th century, high frequency waves were considered less practical for commercial purposes and were thus given over to hobbyist use. These were signals between 3 and 30 MHz, which were called short waves because of their shorter wavelengths compared to the more commonly used lower frequencies with longer wavelengths. Hobbyists and amateurs found the shortwave band had surprising long-distance capability, allowing contact with other amateurs up to many thousands of kilometres away.

On a Monday in May 1960, a few weeks after Francis Gary Powers had been shot down over Russia, Geoff and Derek met at Kettering Grammar School before dawn. The Soviets had published transmission frequencies for Sputnik IV, the first test of the Vostok spacecraft in which cosmonaut Yuri Gagarin would make history the following year. Early Soviet satellites transmitted continuously, so whenever the spacecraft was above the horizon in Kettering, the Doppler shift of the signal would be apparent as the satellite passed by. The *Daily Telegraph* newspaper regularly published sighting opportunities for orbiting satellites—a spacecraft moving across the night sky was a novelty at the time—and Geoff had read that Sputnik IV would pass over southern England at 4:15 that morning.[55]

He later remembered:

> Sputnik IV was launched at midnight on the 15th of May, and I heard about it that Sunday morning [at] 9 on the BBC radio news. I thought, this is what we've been waiting for, so during that morning I cycled up to Derek's and said, 'We've got a new Sputnik up, how about having a go for it tomorrow morning?' Set the alarm for about a quarter to 4, got up, found a pair of trousers, pullover on over the pyjamas, on the bike along Windmill Avenue to the lab. As I cycled to the school, I thought, will Derek be there? Is there too much moonlight? Will we see it? Will it transmit? I'd become accustomed to making visual observations of satellites, looking at them as they went slowly moving, stars across the sky. Would I see it tonight?[56]

In a separate account, Geoff continued, 'At half past four in the morning I was standing outside Kettering Grammar School trying to see Sputnik IV go across the sky, and I failed. But Derek Slater was inside on the top floor with his radio. He picked up the signals and opened the window and shouted something like 'I've got it!' So I forgot all about looking for it and dashed in and there were those bleeps. The first night of trying we were successful.'[57]

Describing listening to Sputnik IV's transmissions through that week, Geoff later recalled, 'Actually, we missed the opportunity to share in a world news scoop. On the Friday, signals were received later than expected. When asked why, I responded that the satellite must have moved higher. It had—unintentionally! The Russians had tried to recover a spherical capsule containing a dummy cosmonaut, but the spaceship had been pointing in the wrong direction when the rocket motor fired.'[58]

Derek also remembered their first tracking session. 'It didn't seem such a big deal at the time. I had spent a year of my life as a wireless operator in the RAF, a lot of it on night shifts. Geoff was quite excited at the unusual thing of getting up at some ungodly hour to hear radio signals, whereas it was pretty routine to me. It was a boost for both of us to hear [Sputnik IV] and it impressed me that Geoff had got the time and frequency right. I knew that, if the signal was there at all, we'd hear it.'[59]

The two teachers made several recordings of Sputnik IV's distinctive bleeping over the next few days. Geoff had finally obtained the Doppler effect demonstration that he had been longing for. The shortwave equipment set-up was dismantled, and Derek's receiver reverted to use by the school's radio club. Kettering's local newspaper featured two articles about the recordings that week, the first news media coverage of Geoff and Derek's satellite tracking activities.[60]

Shortly after making their first recordings, Geoff initiated regular communication with a British government research establishment. In a letter to the UK's Radio Research Station (RRS), he reported a recent unidentified satellite sighting and asked for ground track details related to a Sputnik IV pass. Geoff would provide the RRS with the Kettering Group's tracking results for the next several decades.

The Radio Research Station

The Radio Research Station (originally the Radio Research Board) was a UK government establishment founded in 1920 to investigate properties of the new technology of radio. Some years later, as the Nazis were ascending to power in Germany, RRS scientists were asked by a government agency to invent an electromagnetic ray to shoot down enemy aircraft. They rejected that request but attempted to bounce radio waves off flying aircraft to determine their range and direction. They did so successfully, and the invention evolved into the Chain Home radio location system that helped the Royal Air Force defeat the Luftwaffe during the Battle of Britain. Several other nations developed similar systems in the 1930s, but Chain Home was the world's first operational radar system. The word radar is an acronym for 'radio detection and ranging'—a precise description of the Chain Home network.

In the 1950s, the RRS studied the first satellites launched. A satellite signal receiving antenna, assembled in 1960 in cooperation with NASA, allowed the RRS to track orbiting spacecraft until the antenna system was decommissioned in 1981. Over the years, the RRS distributed prediction information from this data for the orbits of some 1,900 satellites. This prediction information was sent to professional and amateur observers around the UK, including Geoff.

Geoff and Derek decided to reassemble the radio equipment in late 1961 to monitor Discoverer 32, a US satellite—one of the classified CORONA spy missions—that happened to be transmitting a signal their radio could receive. The teachers, of course, did not know about the spacecraft's true purpose. Geoff and Derek listened to Discoverer 32 until its orbit decayed a month later. Using the predicted times of passage over the United Kingdom, they recorded each pass's transmissions and analysed the resulting Doppler shift of the received signal.

Employing John Osborne's method that Geoff had learned in 1957, the teachers determined the point of closest approach to Kettering for overflights of the satellite, and they could thus calculate Discoverer 32's mean orbital period, the time required for exactly one revolution around the Earth. Since a satellite's period is directly related to its height above the Earth, the teachers were able to analyse Discoverer 32's decreasing altitude resulting from atmospheric drag. They plotted a curve of the spacecraft's height loss through to its decay out of orbit.

Discoverer 32 eventually re-entered the Earth's atmosphere at more than 27,000 kilometres per hour (17,000 miles per hour) and burned up.[61] The teachers subsequently listened to Discoverer 36—also a secret US reconnaissance vehicle—from the 1961 Christmas holidays into the New Year. When the satellite's batteries ran out before the mission ended, they used the orbital decay curve they had plotted from Discoverer 32 to predict the date of re-entry for Discoverer 36, which Geoff said they calculated 'with some accuracy.'[62]

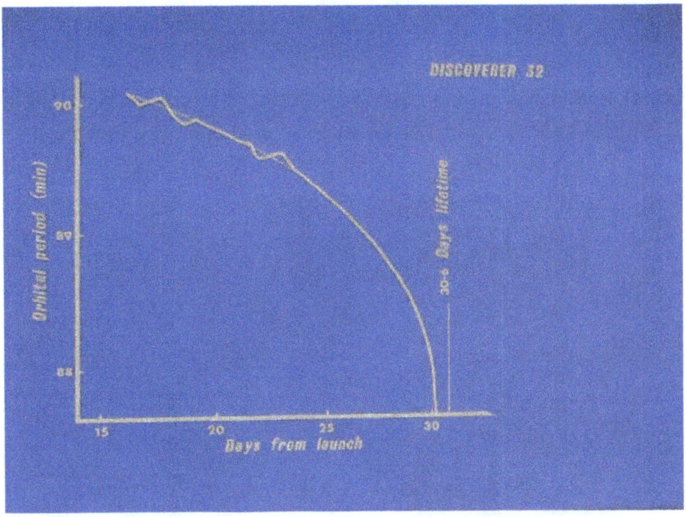

Geoff's chart of the decay of Discoverer 32's orbit, based on Geoff and Derek's Data.

How did Geoff and Derek use World War II-vintage radio equipment to calculate a satellite's orbital period and altitude? Their tools were: a Marconi CR100/2 high frequency receiver; a US Army BC-221 frequency meter; an Advance J-1 audio signal generator; a reel-to-reel tape recorder; an accurate time source, and an oscilloscope.

To record a satellite's signal, they first had to know when a satellite transmitting on a frequency their radio could receive would pass over Kettering. This information was typically available in the daily newspaper near the weather forecast. At least an hour before the predicted pass, the radio and the frequency meter were powered on and their

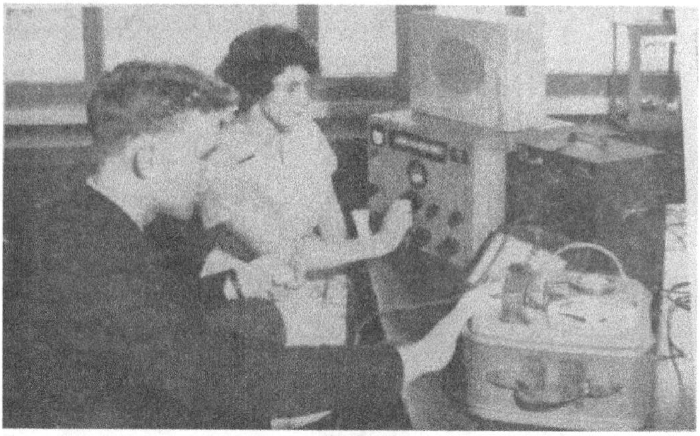

The Kettering Group's original equipment set-up, shown in 1964. The pupils are Christopher Gunstone and Barbara Jacques.

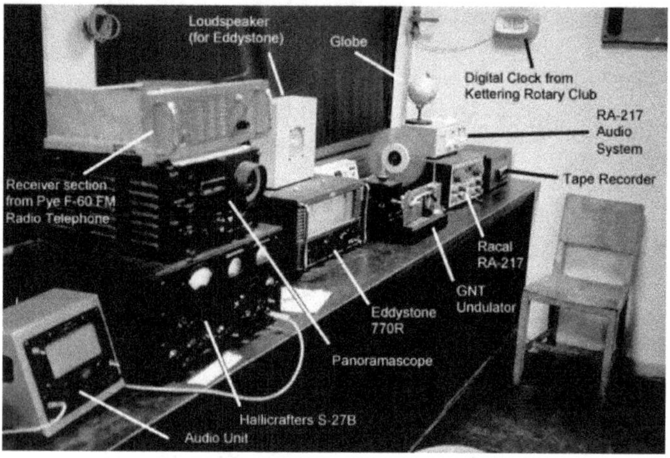

Over the years, the Kettering Group's tracking equipment set-up evolved. This is the Group's tracking station in 1979. The Panoramascope displayed the frequency spectrum of the signal being received. The Eddystone 770R and Hallicrafters S-27B were additional receivers.

valves (vacuum tubes) allowed to warm up. The frequency meter was tuned to the proper setting, which was easily obtained as the Soviet Union publicised their satellites' transmitting frequencies. Then the radio was aligned with the frequency meter's output to enable precise tuning of the signal.

When the received signal was strong, the frequency meter was retuned to just above the satellite's frequency, and the meter's output mixed with the radio's incoming signal. The mixed waves produced a combined signal of a few hundred hertz, which was output to a loudspeaker and recorded by the tape deck. A time signal, typically from a dedicated shortwave radio station, was also put on the tape recording for an accurate reference. Afterwards, the taped output and the output of the audio signal generator were attached to an oscilloscope. The oscilloscope combined the two signals, and the exact frequency of the transmission could be determined by adjusting the controls of the audio generator to achieve a specific pattern on the oscilloscope's screen.

As the recorded signal's frequency changed due to the Doppler shift of the satellite's transmission, the value on the audio generator was noted at specific intervals and the resulting data plotted on a graph. The graph showed the relative change of the satellite's signal frequency with time. By applying mathematical techniques, the time of closest approach of the satellite and the satellite's slant range to the school (the straight-line distance from the school to the location of the satellite at a specific time) could be determined. The time of closest approach was taken over at least two days; a satellite's orbital period could then be determined. The teachers plotted the decreasing period, and thus decreasing altitude, and showed how the satellite's orbit decayed and finally ended.[63] The US' National Security Agency used similar techniques, including Doppler effect analysis, to examine early Soviet satellite transmissions.[64]

Early in the Kettering Group's history, all collected data were analysed manually. In the early 1960s, personal computers and portable electronic calculators didn't exist. Geoff and his pupils initially did all their computations on slide rules and a hand-cranked mechanical adding machine. These calculations were undoubtedly tedious, laborious and time consuming, which was reason enough for Geoff to have his Group members work out the numbers. He taught the pupils how to operate the hand-powered device and was considered an expert in its use. The Kettering Group would use the mechanical calculator until the early 1970s. Over time, the Group's numerical processing methods became more sophisticated.

The two teachers published their method of Doppler analysis of satellite signals in the flagship journal of the British Interplanetary Society, *Spaceflight*, in November 1962. They encouraged readers who made similar observations to send their data to the Radio Research Station at Slough; the RRS staff were always looking to improve the predictions they published by analysing reliable data. A slightly modified version of the *Spaceflight* article was also included as an appendix in the 1963 BIS publication *A Handbook of Astronautics.*[65]

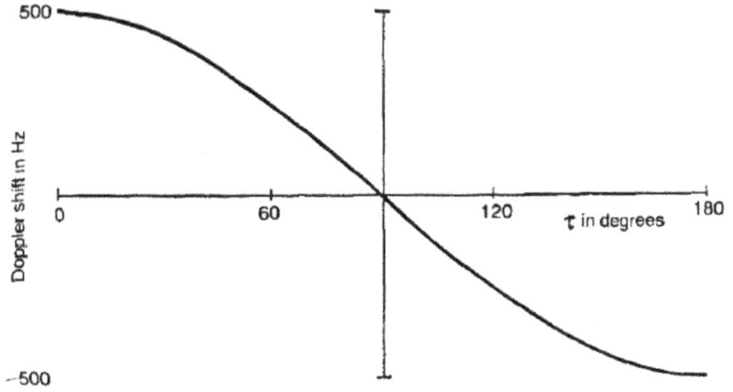

Figure 2 *Graph of Doppler shift against τ for f = 19.995 MHz and v = 7.46 km s⁻¹.*

Doppler frequency shift offset (vertical) v. tau (horizontal) showing Doppler Effect. Tau is the angle between the line of sight to the satellite and the direction of motion. When tau equals 90 degrees, the satellite is at its point of closest approach to the observer, and the Doppler shift is zero.

4

The First Pupil Satellite Trackers

1961 – 1966

Although Geoff's tracking activities initially did not involve his pupils, almost certainly he would have used the headline-making objects circling the Earth to inspire his pupils as to the importance of science and technology. Over time, Geoff began to engage his pupils in observing satellites and this led to an increasing involvement of them in radio tracking, analyses of the emitted signals and studying the satellites' orbital parameters.

Geoff's satellite tracking activities were not limited to radio signal reception. He had read British geophysicist Desmond King-Hele's 1960 book *Satellites and Scientific Research* and was inspired to apply some of the book's ideas to his classroom teaching. King-Hele described how a satellite's changing orbit is connected to the oblateness of the Earth—the flattening of the Earth due to its rotation, resulting in the planet's diameter through the equator being larger than its diameter through the poles. This flattening had been known about and measurable since the late seventeenth century, albeit with great effort.[66]

In his book, King-Hele described a way of calculating the Earth's oblateness that anyone could do with simple tools and a satellite that was regularly visible. Geoff wrote to King-Hele asking permission to use some of his mathematics, and King-Hele replied to approve the request, writing, 'Your method is ingenious and…I think it is a most admirable example for teachers of what can be done with ingenuity and sparse data.'[67] King-Hele became a regular correspondent and information source for Geoff and his pupils.

Since the oblateness of Earth causes a satellite's orbit to precess (see the box below), Geoff could determine the amount of oblateness by measuring the spacecraft's precession and knowing a few basic facts about its path. In 1960, an easily viewed satellite called Echo 1 was launched by the United States. It was a large inflatable sphere about 30 metres (100 feet) in diameter, constructed of Mylar, a thin, shiny material with excellent reflective properties at both radio and optical wavelengths. Echo 1 was used to successfully demonstrate that radio waves could be bounced off the surface of the satellite and sent to a distant receiver. It also vividly reflected sunlight and so was easy to see during morning and evening twilight, when the Sun was below the local horizon but still shining on the spacecraft.

Geoff observed Echo 1 in the spring of 1961 and recorded the changing location of maximum elevation in the sky over numerous orbital passes of the spacecraft. Using the calculations described in King-Hele's book and his own methodology, Geoff demonstrated to his pupils the existence and magnitude of the oblateness of the Earth.[68] 'When Geoff talked about getting space into the curriculum,' Kettering Group member Robert Christy remembered later, 'he was talking about using space as a source of material you could teach with, which nobody had done before.'[69]

Echo 1 satellite inside a hangar before launch.

Precession

If the Earth was a perfect sphere, a satellite launched into orbit would stay in that exact orbit forever (ignoring external forces like friction effects of the atmosphere, solar radiation pressure and the Moon's gravitational force). But the Earth is not a perfect sphere; it has a small bulge—its oblateness—around the equator due to the planet's rotation.

Gravity from the bulge pulls on a satellite and affects its orbit, but not in the way you might expect. The gravitational pull of the Earth's bulge doesn't pull a satellite down towards the equator; instead, the satellite's orbit precesses—the plane the satellite moves in rotates around the centre of the Earth. Even stranger, the direction of the plane's rotation is opposite to the motion of the satellite: if the satellite is moving towards the east in a prograde orbit (a satellite orbiting in the direction of the Earth's rotation), as most do, the satellite's orbital plane rotates towards the west.

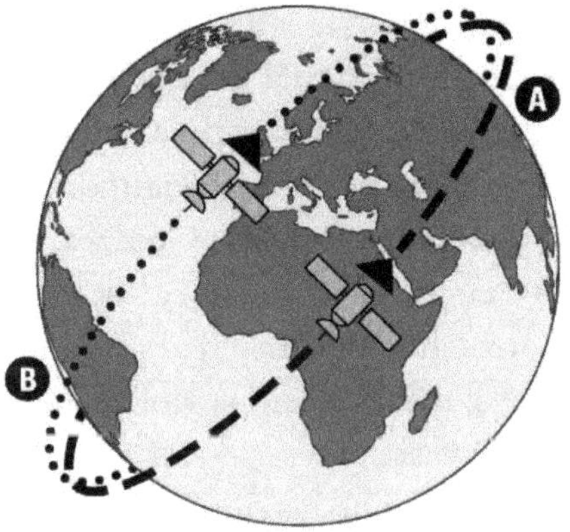

Precession of a satellite's orbit: the prograde orbit of a satellite drifts west over time, from orbit A to orbit B.

Why does a spacecraft's orbit precess? When you apply a force to a spinning object, like a gyroscope, it moves in a direction perpendicular to the applied force due to the rotating motion combined with the force. When an object (a satellite orbiting the Earth) has a force applied to it (the gravitational pull from the Earth's bulge), thesatellite's orbital plane doesn't move down to the equator, but at a 90-degree angle to that motion—in other words, the orbital plane rotates sideways around the Earth's north-south axis.[70]

These observations and classroom activities led Geoff to write his first scholarly article. Published in the 'Physics Notes' section of the November 1961 edition of the British journal School Science Review, the short piece was titled 'To determine the Earth's flattening from the rotation of the orbital plane of an Earth satellite.' This was the first of many letters, articles and papers that Geoff would write over the next four decades.

In the summer of 1961, Geoff attended a physics teachers' course at University College in London. He learned about the satellite prediction bulletins regularly issued by the Radio Research Station, and signed up to receive the predictions, relying upon them to assist his work for many years. The information included precise descriptions of a given spacecraft's trajectory and position across several orbits, which allowed Geoff to determine when a satellite of interest would be passing over the UK so he could listen to it on a shortwave receiver or visually observe it.

In early 1962, Geoff noticed that the Soviets had started launching short-duration missions, mostly under the Cosmos programme name. These satellites would remain in orbit for eight days, making them ideal subjects for Kettering Group projects. Geoff and his pupils started calling these missions 'eight-day wonders' after Desmond King-Hele's label in one of his books. Initially, they had no idea why the spacecraft had such short lifetimes; these missions would intrigue Geoff and his young trackers for many years.

Geoff soon started including his pupils in a visual observation programme. The Echo 2 satellite, launched in January 1964, was another large inflatable balloon, 41 metres (135 feet) in diameter—one-third bigger than its predecessor, Echo 1.

The first Echo satellite didn't cross directly over England because of its low orbital inclination, the angle between a satellite's orbital plane and the equator. Echo 2, though, was in a near polar orbit and so could be seen almost anywhere on Earth when it was above the horizon near sunrise and sunset. Geoff realised that Echo 2 would be easier for his pupils to observe than Echo 1 due to its polar orbit. He devised a simple form for pupils to fill out when they watched for Echo 2 at assigned times and invited all the pupils in the school to contribute observations. By combining numerous reports of the satellite passing overhead, he computed the orbital period of Echo 2 quite accurately.[71] He described an improved method of data processing for this project: 'In 1964 through the friendship of Ron Smith, we used [the computer language] Autocode with the Elliott 803 computer at the CWS corset factory at Desborough to analyse pupils' visual observations of the Echo 2 balloon satellite.'[72] He described his methods and findings in the June 1964 edition of *School Science Review*.

Regarding how pupil involvement first started, Geoff recalled:

> The first pupils to make regular [radio] observations were the sixth form [high school senior] girls from Kettering High School who came to us for physics, chemistry and mathematics and who sat in the empty Physics A [laboratory] during some of their free periods as there was nowhere else for them to go. One day, as I went off to teach in another room, I said to them, 'If that [radio] starts to bleep, note the time for me on the blackboard.' That was the real start of pupil involvement.[73]

From early 1962 until mid-1963, the girls from Kettering High School were the only pupils involved in Geoff's radio observations.

Barbara Jacques, a Kettering High School pupil, monitored the progress of Cosmos 32 in mid-1964 using the Group's equipment. She determined that the last transmission from the satellite occurred during its 128th revolution of the Earth. Jacques concluded that Cosmos 32 had re-entered on that orbit, as there were no further transmissions. Geoff later wrote, '*Flight* [*International* magazine] reported that Cosmos 32 had decayed on 18 June 1964 at 1200 ± 12 hours—no better than saying that it had come down on Thursday! Barbara Jacques and Derek [Slater] had logged it at 1129 BST that day and so I wrote a letter to the editor of *Flight* narrowing down the time, signing it 'Barbara M Jacques, G E Perry & J D Slater'. A letter came back from *Flight* addressed to Miss Jacques and her pupils.'[74]

From left: Geoff Perry, Derek Slater, Barbara Jacques, c. 1963.

While Geoff had exaggerated the size of the decay window *Flight International* had reported—it was about five hours, not twenty-four—this was the first published satellite tracking report written in part by a Kettering Group pupil. The next update of *The RAE Table of Earth Satellites*, the Royal Aircraft Establishment's compilation of data on every spacecraft launch in the world, showed that Cosmos 32's official deorbit time had been 1134 BST on 18 June, just five minutes after Jacques and Slater had last heard the spacecraft.[75] After Geoff straightened out the status of 'Miss Jacques'—being not the teacher but the pupil—*Flight International* sent a reporter to research an article about Geoff, Derek and their pupils' satellite tracking activities.

Jacques was featured in both the *Kettering Evening Telegraph*, Kettering's local newspaper, and the *Flight International* article about the group published the next year and was shown tuning the group's Marconi radio.[76] Another early Kettering Group member from Kettering High School, Suzanne Starmer, said at the time: 'It all started because I had maths and physics lessons up here. And I was asked if I was interested in doing this observation work. I find it very interesting.'[77]

Among the first boys to join the group was David Sinnett. In an interview years later, Geoff recalled: 'I think it was in 1964 that the girls finally got a physics teacher of their own, and I lost my labor. So to take the place of the girls, I invited selected boys to help me with my hobby. One in particular was David Sinnett who lived next door to me at home. At one occasion he was entrusted with the keys to the school on the weekend to do an observation and bring the keys back.'[78]

In October 1964, Geoff had read that a Soviet crewed spaceflight was imminent and tuned the radio to a known frequency. He heard telemetry from Voskhod 1 before the Soviets had announced the flight; this was the first spaceflight in history with three people on board. Geoff phoned the RRS to report details of Voskhod 1's telemetry transmission; the signals sounded much like previous crewed Vostok transmissions. But the official on the phone didn't believe him. 'Mr. Perry telephoned the Radio Research Station at Datchet [near Slough] to report the signals. The expert in the satellite section there suggested that they were from Electron 1. But Electron 1 did not transmit on 19.995 [MHz], Mr. Perry pointed out. Well, maybe it was Cosmos 46, the expert suggested. Since Cosmos 46 was known to have been recovered over a week before, Mr. Perry rang off.'[79]

He had almost scooped the Soviets, but the timeliness of the information hadn't been appreciated. Shortly after the RRS phone call, Geoff spoke with an *Kettering Evening Telegraph* newspaper reporter who told him that TASS had just published Voskhod 1's details and the frequencies it was using to talk to mission controllers. Geoff and his pupils tuned in to hear the voice of Vladimir Komarov, the mission's commander.

Meanwhile, the Kettering Group's radio tracking methods were becoming more sophisticated. As with all observed signals, Geoff and Derek kept a log of shortwave telemetry receptions from Soviet spacecraft and had developed a graphical method of predicting satellite transmission times. On a time versus date graph, they plotted when a specific satellite's transmissions were received. After two or three transmissions, a grid could be constructed that would allow anyone to tune in at a predicted time and hear that satellite.

Geoff and Derek wrote up their method for *Flight International*, and it was published in November 1964 in the magazine's 'Spaceflight and Missiles' news section—indicative of the respect and trust now being placed in the Kettering analyses by the journal.[80] The news item referred to the 'Kettering Grammar School satellite observation group', the first time that Geoff's activities had been given a name in print — though that wasn't quite what Geoff finally decided on: the Kettering Grammar School Satellite Tracking Group.

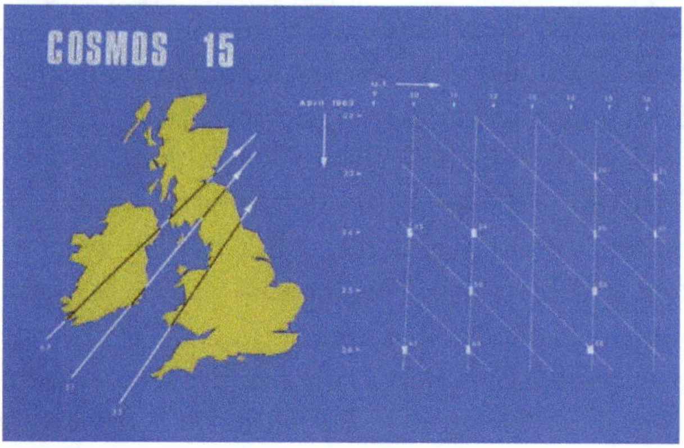

Kettering Group's diamond-shaped grid showing Cosmos 15 ground tracks and listening times. Fellow science master John Marshall made over sixty of these graphics boards for Geoff's presentations.

Geoff and Derek continued to focus their Cosmos observations on missions that were typically eight days in duration. They noticed that on the final orbit of these satellites, radio signals from the satellite started several minutes before the expected time, indicating the deorbit of the satellite was in progress. The lowering of the satellite's altitude resulted in the satellite appearing earlier above the horizon at Kettering compared to if it had remained at its higher altitude; a satellite must travel faster at a lower altitude to remain in orbit. They flagged this observation as a way to identify the end of an 'eight-day wonder' mission.

In May 1965, no doubt encouraged by Geoff, Kettering Group member Richard Jahn published comments on a prior letter to *Flight International* regarding alleged Soviet discovery of the Earth's outer radiation belt.[81] Jahn noted that while the original letter identified Sputnik III as the platform for this claimed discovery, published orbital parameters indicated that Sputnik III did not achieve sufficient altitude to observe the belts. He suggested that several attempted Soviet lunar probes, monitored by US agencies, would have passed through the belt on the way to the moon and certainly could have detected it. Apparently, however, the Soviets intentionally withheld the discovery as they did not publicly acknowledged missions that failed to accomplish their objectives, the Kettering pupil noted.

Astronomer Patrick Moore, well known in the UK for his 'The Sky At Night' BBC television programme, hosted the opening of a space photography exhibition in London in 1967. Already familiar with Geoff's activities through the press and the British Interplanetary Society, Moore invited Geoff and his satellite tracking pupils to tour the exhibit with him.

Richard Jahn later remembered:

> For this function, we were all expected to wear school uniform. Mr. Perry was well aware that most of us planned to make a 'night of it' (as only 17-year-old adolescents let loose in London know how) following the event. Mr. Perry made it clear that we were to wear jackets over our school blazers so that - should we make total drunken idiots of ourselves - Kettering Grammar School would not be identified in any way! What he hadn't counted on was the fact that the event included staff passing amongst the guests with trays of free alcoholic drinks and snacks on little sticks! I still have one of the official souvenir photographs of the event which shows Messrs. Perry and Slater with Patrick Moore surrounded by us schoolboys. Some of us have rather glazed expressions that I suspect were the result of downing far too many free drinks, far too quickly.[82]

The Kettering Group with Patrick Moore at a space photography exhibit in London, 1967.

In keeping with his desire to publish findings and share the Kettering Group's activities, Geoff wrote up a summary of what the Group had accomplished over the previous few years for an education audience. He distinguished between passive and active methods of utilizing current space science to augment physics teaching. Passive methods, he noted, take examples from space science and insert them into the standard curriculum; for instance, adding to a lesson on how a spring oscillates with a report of Lockheed engineers using springs to 'weigh' astronauts in microgravity. Active methods, by contrast, involve pupil engagement in visual and radio observations of satellites to generate data that illustrates physics concepts as well as teaching scientific method.

Along with explanations in the article of the Doppler effect and how high frequency radio signals can bounce off the ionosphere, Geoff described the Faraday effect (see the box below). In the 1840s, famed English scientist Michael Faraday had shown how light waves can be affected when they pass through a magnetic field. Geoff described how shortwave radio signals from Soviet satellites can fade in and out due to this phenomenon and used pen recordings of spacecraft transmissions to illustrate the effect.[83]

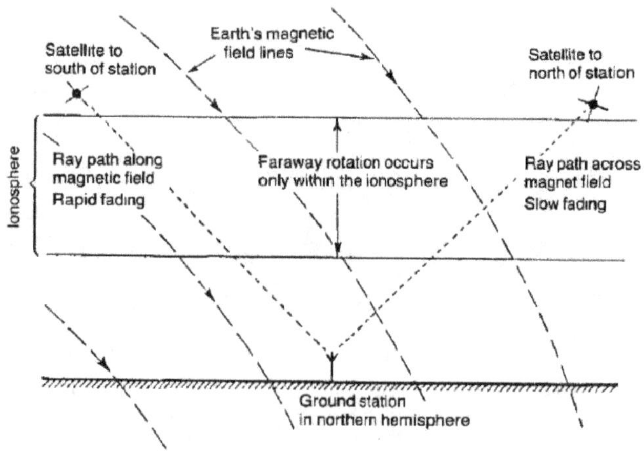

The Faraday effect: signals moving parallel to a magnetic field are affected more than signals moving perpendicular to the field.

The Faraday Effect

Since light and radio waves are both electromagnetic in nature, both can be affected by a magnetic field. When an electromagnetic wave in the ionosphere moves parallel to a magnetic field, the orientation of the wave—whether it oscillates vertically, horizontally or somewhere in between, known as polarization—is changed by the field; this is known as the Faraday effect. The polarization change shows up as a periodic fading of signal strength over time. A radio signal moving perpendicular to a magnetic field through the ionosphere is much less affected.

The signal from a satellite flying overhead might start out with periodic fading, then the fading period would slowly increase as the spacecraft moved away, a result of the angle between the received transmission and the Earth's magnetic field. Geoff's colleague Ken Edwards, at University College of Wales, Aberystwyth, made many pen recordings of satellites in the 1960s that demonstrated this phenomenon.[84]

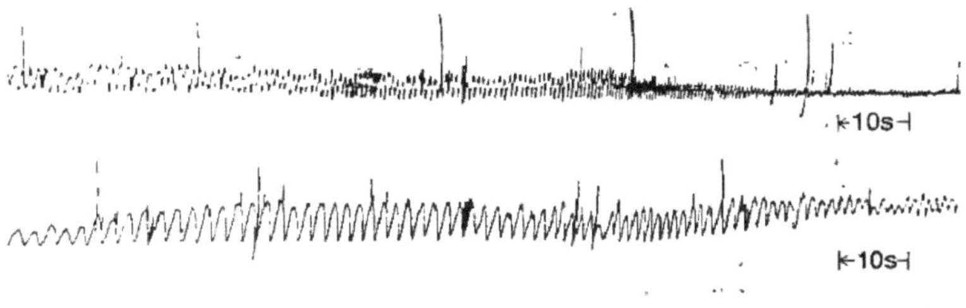

The Faraday effect is evident in this pen recording of Explorer 22 satellite's transmission. The diagram is read from right to left, starting from the top trace. The smaller the width of the trace, the weaker the signal.

Geoff continued his emphasis on writing for education journals to publicise his pupils' activities, and to encourage other instructors to teach creatively. In an article for the BIS journal *Spaceflight*, he summarised the Kettering Group's activities from the earliest observations to the latest projects. Geoff wrote about why he ran the satellite tracking operation: 'I would not like the reader to lose sight of our main reason for maintaining a tracking station at school. This is to provide a rich background from which examples can be drawn to illustrate the traditional school courses in physics.'[85] He also noted that, while the Group had not been used as a recruiting device for physics candidates, many former pupils had gone on to notable studies and careers in the sciences. Geoff concluded, 'One of the most fascinating aspects of a project like this is that it is really 'open ended'. It is impossible to forecast what results will be obtained in the future.'[86] His comment succinctly described both the Kettering Group's satellite tracking activities and the potential course of his pupils' lives.

Geoff explained why satellite tracking appealed to his pupils:

> I think the great thing, the educational value of this was that by inviting them to help me with my hobby, they were being unconsciously taught scientific method. If you say to a British schoolboy 'sit down lad, I'll teach you scientific method', he'll say 'over my dead body, I want to go play football'. You say, 'come help me track satellites', then they come. Over the years, they learn the patience of taking readings, of recording that data systematically, of analysing it, [and] formulating hypotheses to try to explain what they're hearing. When they've got a hypothesis, they try to predict from that hypothesis what should happen. They test it; if it happens, good. If it doesn't happen, you go back and modify. Then, eventually, you publish it. That is what the whole of scientific method is about. They unconsciously learn that over a period of years.[87]

As the Kettering Group's tracking methods developed, more pupil involvement was required. Derek Slater recalled how satellites transmitting on higher frequencies required precise antenna alignment:

> The only thing we lacked was an automatic aerial tracking system, which would keep the beam [of the antenna] pointed at the satellite as it came over our horizon. We solved this one by putting the aerial on the roof of the Science block to be held by one of the tracking group [members]. He was provided with a pair of headphones and

instructed to point the aerial where the signal was loudest....The first trial was on a drizzly afternoon after school. Complaints from the automatic tracking system that the rain was collecting on the aerial and running down his arms were ignored. Later, when a good signal had been recorded for a full six minutes or so, there were congratulations all round. We had completely forgotten the tracking system until there was a plaintive cry from the roof. 'Can I come down yet? I think it's going to thunder.' [88]

The 'automatic tracking system' was pupil David Dean.

David Dean pointing a 'crossed Yagi' directional antenna. The crossed antenna elements minimise signal fading from the Faraday effect.

5

Cosmos 112 and Plesetsk

1966

After a few years of listening to Soviet satellites, Geoff and his pupils knew to expect an 'eight-day wonder' launch about every two weeks. They could predict when transmissions were scheduled, and they made innumerable recordings and log entries tracking the lifespan of their Cosmos targets. But in mid-March 1966, the Cosmos 112 satellite was sounding and behaving differently than previous missions. It would take considerable effort to figure out why Cosmos 112 was unusual.

A few hours after its signal faded at Kettering Grammar School, Cosmos 112 made its first daylight pass over North America, flying up the US east coast. A small group of Soviet spacecraft designers, military technicians and intelligence specialists knew the satellite by its classified name: Zenit (Russian for 'zenith').

The Zenit programme was conceived by Soviet planners shortly before Sputnik I launched in 1957—about the same time that US officials started planning the CORONA reconnaissance satellite system. Zenit was the same type of craft that would later carry the Soviet cosmonaut Yuri Gagarin; it was fitted with cameras, film and supporting equipment. Its primary purpose was to assess American military capabilities and provide accurate targeting information for the Soviet ICBM fleet.

Each pass across the US was carefully planned and instructions were relayed to the satellite in advance via radio link. A Zenit spacecraft initially stayed in orbit for eight days, taking thousands of images of the United States. At first, objects that were 9 metres (30 feet) in size or larger could be resolved in a photograph. By the end of the Zenit programme, photographic resolution had improved to 1 metre (three feet). [89]

Zenit and CORONA

In the late 1950s, the US and Soviet Union leaders independently realised that, if they could put a satellite in orbit, they could install a camera onboard to photograph the other sides' military facilities, equipment and, given appropriate timing, tactical deployments. Both governments authorised secret programmes to develop such spy satellites. Both programmes required extensive development and testing before they were functional. Both programmes started returning useful images in the early 1960s. Both systems were cloaked in secrecy, the US's CORONA spy satellite disguised under the Discoverer label and the Zenit programme of the Soviet Union flying as Cosmos, a catch-all name for their military space missions and many of their science satellites. The rival space powers knew of the other's reconnaissance programme, and both decided not to contest space-based overflight of their territory. The US and Soviet governments realised that CORONA and Zenit were providing vital information that was helping to maintain the uneasy peace of the Cold War.[90]

The descent portion of a Zenit satellite on display at the Russian Medical Military Academy, St. Petersburg, Russia, c. 2007. The round holes are windows for the spacecraft's cameras.

Later in its first day in space, Cosmos 112 flew north over the Midwest United States, where numerous missile launch facilities were located. The satellite would have been sensed by another set of radar antennas, known as the Naval Space Surveillance System (NAVSPASUR) or simply the 'Space Fence'. NAVSPASUR was a US Navy Research Laboratory programme that consisted of a line of transmitter and receiver sites stretched across the southern US which continuously sent a thin fence-like radio signal beam straight up, enabling detection of objects as high as 24,000 kilometres (15,000 miles) above the Earth.[91] The radio signal reflected off satellites as they passed through the beam, and the reflected signal was detected by receivers on either side of the transmitter sites.

Several hours after launch, Cosmos 112 passed over Russia and received updated instructions for its camera operation during the next daylight crossing of the United States. At the same time, the spacecraft flew through the beam of a secret US radar installation near Diyarbakir in eastern Turkey. Designed to monitor Soviet missile tests from the Kapustin Yar launch site in southwest Russia, the radar system at Diyarbakir also had the capability of detecting and tracking objects in orbit.[92] Western intelligence organizations wanted updated information about the path of the satellite to understand its ground track and determine which targets Cosmos 112 was looking at.

Data from the detection of Cosmos 112 at Shemya, NAVSPASUR and Diyarbakir was passed via secure communication line to the Foreign Technology Division's (FTD) operations centre at Wright Patterson Air Force Base in Ohio. Computations by one of the centre's mainframe computers told the FTD analysis team that a new Soviet satellite was in orbit. The resulting information on the newly identified spacecraft was forwarded to the Foreign Missile and Space Analysis Center at CIA headquarters in McLean, Virginia.[93] The CIA's Science and Technology (S&T) Directorate staff noted the launch information and included it in the S&T chief's staff meeting package for the next day.

The morning after Cosmos 112's launch, CIA assistant director Richard Helms chaired a meeting of the agency's senior officers. According to the declassified meeting minutes, the chief of Current Intelligence described events in Indonesia and the upcoming trip of French President de Gaulle to Moscow. The head of Communications detailed several new administrative measures. The staff discussed the situation in South Vietnam (related to the war in Vietnam), prospects there for political stability, and evidence of new missile defence systems in North Vietnam.[94]

Near the end of the meeting, the chief of the Science and Technology Directorate noted a development in the Soviet space programme; the first launch into orbit of a spacecraft from the Plesetsk missile base had been detected the previous day by numerous sensors around the world. Analysts had already identified the spacecraft as a photo-reconnaissance platform, and recent CORONA satellite images confirmed that Plesetsk's facilities had been modified for orbital operations.

The S&T chief also mentioned the expected launch later that day of the latest US reconnaissance satellite, a classified payload called KH-7, from Vandenberg Air Force Base.[95] The 'KH' abbreviation stood for Key Hole, a code name for the all the US' classified photoreconnaissance satellite systems. The KH-7 spacecraft was the successor to the CORONA programme. Codenamed GAMBIT, the KH-7 system delivered high-definition images and was more reliable than CORONA. But the CORONA system would remain in use until 1972 as a proven, effective platform.

At 12:45 pm on the day after Cosmos 112's launch, the Kettering Grammar School pupils heard a faint signal on the radio in the same pattern as the previous day's transmission. Unusually, there had been no announcement of a rocket launch from the Soviet news agency TASS, so the identity of the satellite they were listening to was still a mystery.

The pupils had previously tracked Cosmos 98, whose transmissions had been heard about an hour earlier each day, due to the specific orbit of the spacecraft. Because that day's signal had been received about an hour earlier than the previous day's, Geoff theorised this new satellite was in an orbit similar to that of Cosmos 98, so the transmissions from satellite to control centre might be on a similar schedule. As we've seen, typically the Soviets remotely activated a satellite's transmitter when it was flying over the Soviet Union. The spacecraft would then transmit status information about its systems and receive instructions for its next photography passes over the US.

Two days after being lofted into orbit, the Cosmos 112 satellite on its 31st orbit passed just north of Plesetsk, the site from where it had been launched. Its daylight passes over US—five in the first day, including two over Alaska, and four on day two—would repeat every few days, or nearly so; the repetition would increase the probability of having good weather over the targeted locations, which helped produce useable photographs.[96] At this point in its mission, Cosmos 112 would have taken over 1,000 photographs of military facilities and potential nuclear weapons targets in the United States.

Satellite Ground Tracks

Geoff used the motion of satellites to illustrate several principles of physics. For instance, once a satellite is inserted into orbit by its launcher, it moves steadily around the Earth at a fixed angle to the equator. The Earth turns west-to-east under the satellite's orbit, resulting in the spacecraft flying over different parts of the globe on each orbit.

If a satellite's orbit takes 1.5 hours, it will fly 16 orbits in 24 hours, then closely repeat its ground track from the previous day. But most satellites in low Earth orbit take a bit more or less than 90 minutes per orbit. So, after 24 hours the satellite's track might be shifted west or east compared to the corresponding track from the day before. The time to fly one orbit, called the orbital period, can be used to compute future ground tracks and to know when the spacecraft will be overhead a particular location.

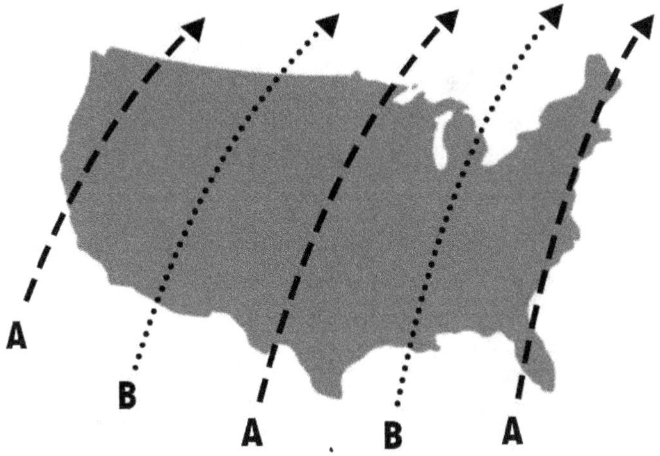

Cosmos 112 ground tracks across the continental US, 17 March to 25 March 1966. The satellite flew along tracks labelled A on flight days 1, 3, 5 and 7, and tracks labelled B on flight days 2, 4, 6 and 8.

On most Saturday mornings, Kettering pupils who had behaved badly during the week served a detention period at the school. A former pupil recalled, 'As he was staff, Perry had to supervise at Saturday morning detentions. He would set long multiplication sums on these occasions and take delight in demonstrating how quickly he could calculate the answer 'on my little machine here (Geoff's mechanical calculator).''[97] Geoff would entertain his captive audience and teach them at the same time. If he wasn't supervising errant pupils, he would often spend part of his weekend in the physics classroom catching up on paperwork and preparing for the following week's classes. In either case, he would turn the shortwave radio on to listen for signals.

Isabel, who was 10 years old when Cosmos 112 flew, remembered, 'It was a very natural thing for me to go up [to the school] at weekends. [My father] would have been listening, developing black and white photographs in the dark room at school, doing some schoolwork. I would go along and do some soldering and weigh old pennies on the new electronic balance. It was just all part of my life. The radio would be on at the back of the lab and you'd track.' [98]

Geoff Perry and daughter Isabel in the Physics A laboratory, Kettering Grammar School, c. 1967

On the Saturday after Cosmos 112's launch, Geoff expected to hear from the mystery satellite about an hour earlier than he and the pupils had the previous day. For the third day in a row, weak signals were heard on the standard Cosmos frequency, this time at about 11:45 am, as was predicted. Geoff assumed the poor reception was probably due to the school's main receiver having failed, and the backup that was now in place not being up to the job. Fortunately, the Marconi radio was repaired, presumably by Derek, and back in operation the next school day.

During the regular lunch hour monitoring on the Monday, signals were heard once more, this time loud and clear. From 12:45 pm to 12:49 pm, a distinctive bleeping was heard throughout the Physics A laboratory, but it turned out to be a transmission from the next launch, Cosmos 113, which had occurred that morning.[99] For each signal detected, whether it was identified or not, a pupil would write down the time, another would ensure the tape recorder was on, and they would note the time again as the signal faded. One of the habits Geoff Perry was teaching his pupils was precise data gathering. Ferreting out clues pertaining to satellite behaviour and purpose required one to identify patterns in the data. The larger amount of, and the more accurate, the data gathered, the more meaningful the resulting information.

On the Tuesday following the launch of Cosmos 112, Geoff opened a letter from Desmond King-Hele. Geophysicist King-Hele was also an expert in satellite orbits and the science that could be derived from them. He had written two books about satellites that Geoff had absorbed and was inspired by: *Satellites and Scientific Research*, noted earlier, and *Observing Earth Satellites*, which chronicled methods of determining satellite orbits and briefly described Geoff's activities.

Geoff subsequently wrote a review of King-Hele's book *Observing Earth Satellites* and summarised it: 'If [the book] has a message, it is that more and more reliable observations are continually required, especially of the short-lived low perigee satellites—and these are among the easiest to observe.'[100] King-Hele gave a shout-out to the Kettering trackers in a section describing how to listen in to spacecraft transmissions on shortwave. He updated the book in 1983 and included several more mentions of Geoff and his pupils in the new edition. Especially in the 1960s, King-Hele regularly gave the Kettering Group data and encouragement in their satellite tracking activities.

Geoff had phoned King-Hele about Cosmos 112 on the day it was launched (though he didn't know the satellite's designation at the time), discussing a few unusual characteristics of the spacecraft and its transmissions. The letter from King-Hele concerned details about the new satellite, which he identified as Cosmos 112. King-Hele's note contained a crucial bit of information: the satellite's inclination to the equator was 72 degrees.[101] This inclination had never been used before for a Soviet satellite.

About Cosmos 112, Geoff said later, 'We picked up signals that I said were "unusual, very weak amid much noise"[...] So obviously here was something quite new.'[102] And he wrote later, of a different Plesetsk-launched mission: 'Signals were once again of short duration, weak and possessed a certain "throaty" quality—like it wanted to clear its throat—now recognised as characteristic of signals from satellites commanded from the northern [Plesetsk] site.'[103] The weak, throaty signals may have resulted from the satellite being farther away from the UK when Soviet controllers sent the command to turn the spacecraft's transmitter on.

Later that day, Geoff received another letter at work, this one from Sven Grahn. Grahn had previously sent a recording of what he thought was a satellite to Geoff. Geoff replied, confirming that Grahn had recorded signals from Cosmos 104 and encouraged him to keep listening while making accurate records. Sven heard Cosmos 112's signals and concluded in his letter that 'Cosmos 112 is a very puzzling satellite', having also noticed its unusual transmission characteristics.[104] After the first successful reception that day of signals at 12:45 pm, the next day Geoff and his pupils expected to hear from Cosmos 112 at about 11:45 am. They monitored the frequency until noon, but no transmissions were detected. Normally a lack of signals might indicate that the satellite had deorbited, but as Geoff said in a reply letter to King-Hele, '[T]his one has been much too unusual to make any dogmatic statements of that sort!' [105]

A Zenit satellite being processed for launch, c. 1968.

Construction of the Plesetsk missile test range officially started on July 15, 1957, in total secrecy. For many years, the Soviets didn't refer to Plesetsk by that name; they gave it the code name Angara. Belgian space historian Bart Hendrickx wrote, 'In an attempt to cover up its real purpose, Angara was [subsequently] named the "3rd Training Artillery Firing Range" (3 UAP) on 7 February 1959 (although the name Angara continued to be used, even in official documentation). In order to keep up the facade, artillery tests were regularly conducted at the range, with local residents being warned not to enter off-limits areas.'[106] In 1963, Plesetsk was again renamed as the Scientific Research and Test Firing Range No. 53 (NIIP-53), continuing the Soviet habit of secrecy by obscuration.

Despite difficult construction conditions, due to the remoteness of the site and challenging northern Russia weather, builders of the Plesetsk missile site persevered. After a successful test of the R-7 missile in late October 1959 by a Plesetsk missile crew at the Baikonur site, the first operational ICBM launch pad in the USSR was declared ready for use on 15 December 1959 at the Plesetsk facility. A second pad followed on 17 February 1960.[107] For the first time, the Soviets could launch an R-7 missile topped with a nuclear warhead as far as 8,000 kilometres (5,000 miles). Washington, D.C. was 7,400 kilometres (4,600 miles) from Plesetsk, and New York was 7,000 kilometres (4,400 miles)—well within range of the USSR's rockets.

US intelligence experts desperately wanted confirmation of possible missile base construction activity near Plesetsk. The first successful CORONA photography mission in August 1960 did not fly over Plesetsk, but the next one to return images did. In December 1960, Discoverer 18/CORONA mission 9013 detected 'an extensive area of activity and construction encompassing approximately 45 square miles [117 square kilometres] and served by rail and road… located in heavily wooded terrain 15 [nautical miles (28 kilometres) northeast] of Plesetsk.'[108] A 3.4 kilometre (2 miles) long runway about 48 kilometres (30 miles) from the newly discovered construction was also noted. But the quality of the pictures was not sharp enough for CIA to positively identify missile launch facilities. In August 1961, Discoverer 29 returned higher quality images of the Plesetsk area, all but confirming the existence of ICBM facilities there.[109]

The following May, the US Department of Commerce published the first unclassified US government study of the Soviet Union's space programme. The report described the robotic and human space missions flown by the Soviets through early 1961, and the technology employed in those flights. Curiously, the author of the study did not include discussion of anticipated or possible military uses of the Soviets' rocket technology. However, the CIA published a top-secret report on Soviet missiles and spaceflight operations the same year and correctly noted that, at the time of the analysis, the Soviets had not launched any spacecraft with overtly military objectives.[110] The Commerce Department's unclassified study was the first of a series of US government open-source reports on the Soviet Union's space efforts. By the late 1960s, those reports would include data and analyses from Geoff and his pupils.

Throughout 1961, the CIA was still working to identify the construction activity near the village of Plesetsk. In June, a classified National Intelligence Estimate (NIE) report said that Plesetsk was 'suspected' of being an ICBM base that had been operational since January 1960.[111] CORONA missions during the summer of 1961 provided more evidence that Plesetsk had an ICBM capability, though this was still not conclusive. The NIE official estimate that autumn was more definitive, saying, 'Plesetsk is an ICBM complex with rail-served launchers designed to employ the first-generation ICBM.'[112] In 1963, after subsequent photo-reconnaissance missions provided more details about the complex, the CIA positively identified multiple launch pads at Plesetsk for all three types of then-known operational Soviet ICBM missiles but did not detect any facilities that could have sent satellites into orbit.

The Soviets started launching the Cosmos satellite series in March 1962. The first few were for testing and research, but Cosmos 4, allegedly a scientific spacecraft, was in fact the first photo-reconnaissance mission in the Zenit programme. The images from Cosmos 4 were unusable, but the next reconnaissance flight, Cosmos 7, demonstrated that the Zenit cameras were viable.[113] Zenit testing continued through Cosmos 20, whereupon the system was declared operational. For the next few years, almost every second spacecraft labelled Cosmos was a spy satellite, averaging nearly two missions a month.

Not all the initial Zenit missions were successful, though. Early in the programme, twice within 12 months, a modified R-7 booster with a Zenit payload exploded on the pad during launch preparations at the Baikonur complex.[114] However, the Zenit programme was quite successful overall: according to one of its system designers, the initial Zenit images 'exceeded even the most optimistic expectations of the client'—the client, most probably, being both the KGB, the Soviet Union's functional equivalent to the CIA, and the GRU, their military intelligence organization similar to the US' Defense Intelligence Agency.[115]

Since the mid-1950s, US intelligence had been routinely monitoring Soviet launches. Using radar and listening posts around the world, the Western intelligence community gathered information on Soviet spacecraft and combined that with Soviet launch announcements to deduce the satellites' purposes. However, as late as December 1962, after the first few Cosmos-labelled test flights of the new reconnaissance system, the CIA was still unaware of Cosmos' photographic abilities.

In late 1964, the first US classified report on Soviet photo-reconnaissance satellites was published. The CIA concluded that the Soviets were photographing the United States primarily for ICBM missile targeting. The report noted that the Soviets were interested in 'a tacit understanding with the US on reconnaissance satellites'—that both sides knew the other was photographing their adversaries' military facilities for offensive purposes.[116] But as late as 1965, the US intelligence community had no evidence that the Soviets were preparing a third orbital launch facility. At that time, all the Cosmos missions had been sent into orbit from the Baikonur and Kapustin Yar launch sites.

While Plesetsk had been identified in classified reports as an ICBM base, it had yet to be recognised as a potential satellite launch facility.

Certainly, the modifications to at least one launch pad for orbital capability at Plesetsk were in progress by 1965. Perhaps images from the early CORONA missions weren't capable of the resolution needed to see such changes. However, a 1964 CORONA image of Plesetsk did document multiple ICBM launch pads—at least 13—and reinforced the notion that Plesetsk was an active ballistic missile base.[117] Since it was only in the summer of 1963 that the Soviet leadership authorised the use of Plesetsk for orbital launches, the construction and upgrades enabling satellite missions wouldn't be complete until 1966, when Cosmos 112 launched.

As the R-7 missile was making its debut as the Soviet Union's main ICBM, plans were already in place to replace it with a rocket that was easier to store and launch. The R-16 missile was being developed to use solid fuel and to be ready to launch with 15 minutes' notice, as opposed to the R-7's liquid propellants and hours-long preparation time. After the US announced plans to place the new Minuteman ICBMs in underground silos, the Soviets started planning to use silos for their missiles. When the R-16 had its first sub-orbital test launch in October 1963, it was the first missile launch of any kind from Plesetsk, and the first Soviet missile launched from an underground facility.

According to space historian Bart Hendrickx, 'In order to mask the [Plesetsk] facilities, the pads had been covered with 'horizontal and vertical camouflage nets' and much of the support infrastructure (missile, propellant and warhead storage facilities) had been built half underground, making it hard to detect amid the taiga [pine and birch] forests. However, the best way to hide the missiles was still to build silos. These would have to be 50 metres [165 feet] deep and have a diameter of up to 30 metres [98 feet].' By late 1962, eight ICBM launch pads were fully operational at Plesetsk, four for the older R-7 and four for the new R-16 missile.[118]

Meanwhile, in the early 1960s, Soviet space planners decided they needed another satellite launch site to supplement the Baikonur and Kapustin Yar facilities. With multiple new ICBM projects progressing, they also needed another missile test facility. In 1961 a military had committee recommended building an orbital launch site in the northern latitudes to accommodate higher inclination missions for both scientific and military purposes. After an abortive attempt to construct a launch base on swampy ground south of Plesetsk, in September 1963 an official decree authorised the modification of ICBM launch pads at the existing Plesetsk complex for orbital missions. Plesetsk also was designated as a test facility for new and modified ICBM missiles, along with its operational ICBM mission. Occasional sub-orbital ICBM launches from Plesetsk sent dummy warheads across Russia to a test range in the far east of the country on the Kamchatka peninsula.[119]

On 25 March, the eighth day of its mission, Cosmos 112 had expended its film supply and the satellite's rocket engine was commanded to fire for re-entry. Unlike the US' CORONA system, which ejected a separate film container for re-entry and recovery while the satellite continued in orbit, the entire Zenit module containing cameras and related mechanisms was deorbited to retrieve its exposed frames. After braking parachutes deployed, a radio beacon signal announced the location of the descending satellite; Kettering Group members would later discover unique characteristics of this signal during Cosmos missions. Recovery forces in helicopters homed in on the beacon and landed next to the scorched spacecraft. Its valuable photographic payload was removed and sent to a special processing laboratory for development and analysis.[120]

The day Cosmos 112 landed, the US' North American Air Defense Command (NORAD) published the *NORAD Weekly Intelligence Report* containing two articles about Plesetsk and a story about the unexpected launch of Cosmos 113.[121] In the since-declassified report, the first Plesetsk article summarised the mission of Cosmos 112. The story explained that the discovery of the new launch site was determined by the 'backtracking of Cosmos 112's orbit' to the estimated time of launch, which put the satellite near the Plesetsk missile test range—the same analytical method that Geoff and his pupils would eventually employ. Unusual characteristics of the mission were described, such as the 72-degree inclination, as Geoff and his pupil analysts had noticed.

Also mentioned were 'telemetry indications of camera-monitoring activity' indicating a photo-reconnaissance function. Geoff and the Group would later decode Cosmos telemetry to determine satellite functions and purpose, similar to NORAD's analysis.

The second article in the report described the expected uses of the missile range's new orbital capability. Since Plesetsk is located so far north, the article speculated that polar orbit missions, mainly civilian in purpose, could be launched from the site. The continued use of Plesetsk as an ICBM training site clearly implied a dual-use for the facility: sub-orbital missile testing and Earth-orbiting spacecraft launches.

A third article, discussing the subsequent Cosmos 113 launch from Baikonur that week, noted that Cosmos 112 and 113 were in orbit simultaneously. Based on prior Soviet photo-reconnaissance mission schedules, analysts considered this unusual. The article described what appeared to be a delayed launch of Cosmos 112, compared to the expected launch date—a type of pattern recognition also used by Kettering Group members. In fact, the Soviets had tried to launch Cosmos 112 earlier in the month but had encountered technical issues for several days while working to get their first Plesetsk-based satellite into orbit.

In early April 1966, just after Cosmos 112 landed, Geoff mailed two letters, one to new Group member Sven Grahn in Sweden, and the other to Desmond King-Hele at RAE Farnborough. Geoff had received the Radio and Space Research Station prediction bulletin (the Radio Research Station having been renamed the RSRS) for Cosmos 112 and a few other satellites some days earlier. He wrote to both correspondents that, with the help of his pupils, he had determined that Cosmos 112 could not have originated from Baikonur, the Soviet launch site in Kazakhstan. Two clues drove this conclusion.

The Cosmos 112 prediction bulletin from RSRS, received on 25 March, provided data on the satellite's orbit. With the help of a mainframe computer program he had designed, Geoff could draw a ground track of the spacecraft's first orbit (labelled revolution 0 in the prediction bulletin). Given the satellite's estimated lift-off time, the revolution 0 ground track Geoff had drawn showed that the launch point was nowhere near Baikonur. Then, the monthly update to the *RAE Table of Earth Satellites*, arrived at the school a few days later. According to the update, Cosmos 112's lifetime was 7.79 days, which equated to 122 revolutions, less than the 128 orbits of the normal eight-day Cosmos mission, implying that either the launch site or the landing site, or both, were different from the usual locations. Geoff suspected that it was the launch site that was different, almost certainly because of the revolution 0 ground track he had computed.

Replying by mail a few days later, King-Hele said that Geoff had good reason to suspect a new launching site but that pinpointing it might be difficult. (King-Hele probably had knowledge of Plesetsk from classified sources, since he held a security clearance.) In his letter to Sven Grahn, Geoff had included a diagram of Cosmos 112's initial ground track. Grahn suggested three possible launch sites along that track. The southern tip of Novaya Zemlya Island, the USSR's nuclear testing site in the Arctic, seemed the most likely of the three because the location was associated with known Soviet military activity.[122]

Geoff had two methods of calculating ground tracks: calculations based on observations by the Group, and satellite positions provided by RSRS information. The first relied on the Doppler shift observations (as described earlier, for greater detail see Appendix B), along with the orbital inclination announced by TASS or obtained from the RSRS bulletin. Given these data, calculation of the launch time and landing time could be made, as well as computing the satellite's location throughout its mission.

Geoff created an algorithm for calculating a satellite's ground track using this information. In a typical Geoff-like manner, this algorithm relied solely on the fundamental principles of satellite orbits and the keen observations made by him and his pupils. At first, the pupils ploughed through the mathematics of the algorithm with a mechanical calculator. Then Geoff acquired access to a mainframe computer in town and, with some help, wrote a computer program to accomplish in seconds what would otherwise have taken hours of manual work on the calculator. A detailed summary of the algorithm for this program is provided in Appendix C.

The second method Geoff used to calculate satellite tracks involved the RSRS prediction of satellite positions. A prediction bulletin specific to one satellite consisted of the basic orbital parameters for the satellite, equator crossing times and longitudes for several orbits, and crossing times at various latitudes for a specific orbit. With this data, Geoff could construct a ground track for any orbit included in the bulletin.

In mid-April 1966, a second orbital launch, Cosmos 114, lifted off from Launch Complex 41 at the Plesetsk site. It was blandly described by TASS as 'carrying scientific instruments', but it actually was another secret reconnaissance spacecraft. Flight controllers would have tested the satellite's functions prior to its first pass over the United States. Cosmos 114's orbital parameters and mission lifetime were similar to Cosmos 112. The second Zenit mission launched from Plesetsk continued the process of gathering information on America's military forces.

A note from Desmond King-Hele confirmed the new spacecraft's orbital inclination at 73 degrees. That, plus the weak signals resembling those previously received during Cosmos 112's flight, suggested to Geoff that Cosmos 114 had also been launched from the new northern site. But since the inclination of Cosmos 114 was nearly parallel to that of Cosmos 112, he couldn't find a clear intersection point of their revolution 0 tracks that could pinpoint the location from where they had both been launched.

On 12 April 1966, Geoff composed a letter to *Flight International.* He described the unique characteristics of Cosmos 112's trajectory, comparing it to a previous launch from the Baikonur complex, Cosmos 98. He concluded that, because Cosmos 112 had ended its mission early, either the launch site or the landing site was not the same as the earlier mission. Geoff wrote that due to peculiar characteristics of the radio transmissions received at Kettering, he suspected that the launch site was different, probably farther north than the normal facility used.

For some reason he made no mention of his calculation of the satellite's revolution 0 ground track. He did note that the recent launch of Cosmos 114 had had a nearly identical trajectory to Cosmos 112, so it probably had also been launched from this new northern site. He credited Sven Grahn for suggesting the Novaya Zemlya launch site possibility. Geoff's letter was published in the magazine's 21 April 1966 edition.[123] To his surprise, there was no public reaction from British or American authorities—possibly because they already knew about the Plesetsk site, and they didn't want to reveal that knowledge.

Although Geoff claimed that he and his pupils were the first to announce the previously unknown third Soviet launch site in the open press, an Associated Press article dated 21 March 1966, a month before Geoff's public announcement, cited a West Berlin observatory source claiming that Cosmos 112 was launched from the vicinity of Novaya Zemlya, as Geoff had first surmised.[124] However, the German observatory apparently never narrowed down the true location of the new launch site, which turned out to be Plesetsk, more than 1,200 kilometres (746 miles) southwest of the southern tip of Novaya Zemlya.

The third orbital spacecraft launched from Plesetsk was Cosmos 121, yet another 'eight-day wonder'. The Soviets were reliably sending two such missions into orbit every month (though not all from Plesetsk) to the delight of the Kettering pupils who discovered and tracked them. This spacecraft's orbit, like the first two satellites launched from Plesetsk, had an angle to the equator of about 72 degrees. Frustratingly, it also could not be plotted against the previous tracks of Cosmos 112 and Cosmos 114 to obtain a crossing point that would identify the launch location.

But Cosmos 121 did provide another clue pointing to a possible new launch site. One of Geoff's pupils heard a faint Cosmos-style signal just after noon the day of Cosmos 121's launch, the radio having been otherwise quiet that morning. Several similar transmissions were heard throughout that day. Typically, the first audible signal of a new spacecraft was picked up at Kettering at the end of the satellite's first orbit. When a transmission from what was assumed to be the last orbit of Cosmos 121 was heard at the end of the eight-day mission, the pupils were surprised to hear signals again 90 minutes later.

Upon receiving the *RAE Table of Earth Satellites* update for Cosmos 121, they realised that the first transmission they had received on launch day had occurred shortly after lift-off of the satellite, not at the end of its first orbit. Receiving signals from the satellite as it ascended into orbit suggested a launch location closer to Kettering, as signals from satellites being launched from Baikonur had never been received during the ascent phase of flight.[125]

On 14 October 1966, a fourth orbital mission was launched from Plesetsk. Cosmos 129 was the key to revealing what Geoff Perry and the Kettering Grammar School Satellite Tracking Group was looking for. This spacecraft, the 41st of 82 Zenit reconnaissance missions to be launched, was inserted into an orbit that differed noticeably from the previous Plesetsk missions. The first three had flown at inclinations close to 72 degrees, so their orbital track intersection could not definitively show where their launch site was. But Cosmos 129 was flying at a sufficiently different inclination so that its revolution 0 ground track would cross the others at an angle that could reveal the common point of origin.

Following the launch of Cosmos 129, Geoff and his pupils monitored 19.995 MHz, the standard frequency that Cosmos satellites transmitted on. They picked up signals, but the reception was brief and during the next several days the transmissions were distorted and short-lived, much like those of the previous missions that were thought to have originated from the new launch site. TASS had announced the mission of Cosmos 129, so the pupils assumed that was the satellite they were hearing.

Geoff contacted the RSRS and learned that Cosmos 129's orbital inclination was near 65 degrees. This angle to the equator was typically associated with missions launched from Baikonur. But the odd qualities of the received signals, plus the relatively late launch time provided by RSRS, suggested that Cosmos 129 might have originated from the new launch site. Another anomaly the Group later uncovered was the early landing during the seventh day of the mission. To accommodate such a recovery time, the initial orbital ground track for Cosmos 129 could not line up with Baikonur—it had to have followed a ground track further to the west.

Two days after Cosmos 129 landed, Geoff asked the RSRS for a prediction bulletin for the spacecraft. Using that information, he determined the ground track of Cosmos 129's first orbit. Plotting the track, along with those of the previous missions thought to have been launched from the northerly site, the revolution 0 track lines converged at a point about 162 kilometres (100 miles) south of Archangelsk, a city on the edge of the White Sea in Russia's northwest. He calculated the launch location to be latitude 63 degrees north, longitude 41 degrees east, with a potential error of plus or minus one degree for each measurement.[126] Geoff's calculations were remarkably accurate: his prediction was within half a degree in longitude and within a tenth of degree in latitude of the actual launch site position.

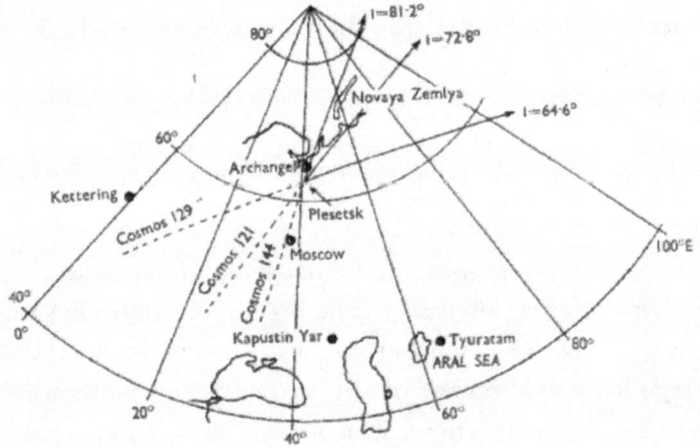

The intersection of Cosmos ground tracks plotted by Geoff revealed the Plesetsk launch site.

Geoff was keen to announce his new finding publicly; he found the opportunity at the Autumn Meeting of the British Interplanetary

Society in late October 1966. Founded in 1933, the BIS is the oldest extant space advocacy group in the world. Geoff had been active with the BIS for several years and had been regularly publishing his results in their flagship journal, *Spaceflight*. At the meeting, he gave a talk on 'Radio Observations of Satellites' as part of a series of presentations on the use of visual aids in astronautics education. During the talk, he revealed the coordinates of the previously secret Soviet orbital launch site. His announcement was met with little fanfare.

Geoff also sent a letter to *Flight International* with the new information, which was published a week later.[127] He was again surprised by the absence of reaction from authorities and the press, as he had been by the lack of attention to his April letter describing the preliminary results of his Cosmos 112 analysis. Pupil Michael Sinnett later recalled the dismay from the apparent lack of interest in the Kettering Group's Plesetsk discovery. 'The atmosphere in the Group was very low at that point, because we felt that here was something that we'd really achieved. We'd been recording information for a long, long time, for years, and we'd finally hit upon something that either no one else knew about or at least they hadn't told the world about. We'd found it, we were ready to tell the world about it and then the world didn't seem to want to know.'[128]

That dearth of interest, however, was not to last.

The United States government had a keen interest in determining what the USSR was doing in space and what its goals there were. Analyst Charles Sheldon, chief of the Science Policy Research Division in the US' Congressional Research Service, had been tasked in 1966 with writing an unclassified review of the Soviet space programme for the US House of Representatives, which became an annual publication. He was putting the finishing touches on the report and intended to send it to several government agencies that needed to approve it for accuracy and security.

Sheldon wanted to describe the newly revealed northern Soviet launch site—he had seen Geoff's letter in *Flight International*—but its existence was considered classified information in the US. Because he held a security clearance, Sheldon presumably knew of Plesetsk and the activities there. But his report to Congress had to be based solely on verified sources in the public domain.

Congressional Research Service

The Congressional Research Service (CRS) is a unit of the US Library of Congress. Established in 1914 as the Legislative Reference Service, it transformed into the Congressional Research Service in 1970, with significantly expanded responsibilities. The CRS provides research and analysis services to the House of Representatives and the Senate for all topics and issues of interest to the legislative bodies. The CRS published numerous studies of the Soviet space programme from 1967 until the dissolution of the Soviet Union in 1991 and has continued evaluating the USSR's space programme. Geoff and his pupils contributed to many of these studies, as we shall see later.

Sheldon wanted to use the information about Plesetsk in his upcoming report to the House. He knew the CIA considered the existence of Plesetsk classified and would probably not be impressed by the site being revealed by a schoolteacher and his pupils. But he was acquainted with a few journalists who might be interested in verifying the story. If a major US press outlet printed the information, he hoped that the CIA would let him include it in his report.

In late November 1966, Sheldon had apparently talked to his friend Joseph Myler from United Press International and asked Myler if he would follow up on the *Flight International* article about the new Soviet launch site. Myler's reporting evidently confirmed that Geoff and his pupils were trusted in the British aerospace community and that their activities had been vetted by *Flight International* editors. In a 13 December 1966 *Washington Post* front page article, Myler referred to anonymous 'US sources' confirming the existence of 'a big military base south of Archangelsk'.[129]

Alfred Friendly, a London-based correspondent for the *Washington Post*, saw Joseph Myler's story and wanted to write a more detailed follow-up piece on Geoff and the Kettering satellite trackers. Friendly, along with two writers for *Time* magazine who had also caught wind of the story, interviewed Geoff and several pupils and was given a demonstration of their satellite tracking methods. Friendly's page one report in the 19 December 1966 edition of the *Washington Post* was the first detailed account of the Kettering Grammar School Satellite Tracking Group and its activities in a US publication.[130] Friendly included in the story details of how Geoff and the Group had pinned down the Plesetsk launch site.

During his many years in Washington, D.C., Charles Sheldon had developed a large network of colleagues. One of his acquaintances was Philip Klass, senior editor at the magazine *Aviation Week and Space Technology.* Sheldon apparently also spoke to Klass about Geoff's activities. On the same day as Alfred Friendly's story was published in the *Washington Post*, Klass wrote a piece for *Aviation Week* analysing Plesetsk's function as a possible location for defending against any US-launched missiles headed for Russia. Klass, like Joseph Myler, didn't use the name Plesetsk, referring to the launch site as 'near Archangelsk'.[131] In the same issue, *Aviation Week's* London bureau chief Herbert Coleman wrote a separate article about Geoff and his pupils' activities, describing the methodology behind their use of the Doppler effect for satellite tracking.[132]

With newspaper and magazine headlines starting to proclaim the Kettering Group's achievements, press enquiries streamed into Kettering Grammar School and Geoff's life was in an uproar. Although the school term had ended the previous day for the Christmas holiday, reporters were clamouring for information. *The Times* noted Alfred Friendly's report in the *Washington Post,* suitably impressed that the Kettering trackers had made their observations and analyses with 'Army surplus equipment and a toy globe from the local stationery shop'.[133]

A British Pathé film crew captured the pupils hard at work tracking a satellite and recording data. One scene from the film shows a map with two satellite ground traces on it and the latitude and longitude of the point where the two tracks cross—it appears to be the actual drawing that pinpointed the existence of the secret launch site.[134] Geoff later remembered, 'The day after we broke up for the Christmas holiday, the whole of the world's media descended upon Kettering. There were sixty phone calls, five TV crews, they got me out of the barber's chair, out of the bath, they stopped me playing golf. I was so excited it upset my stomach. I had nothing alcoholic at Grandma's at Christmas and I've been teetotal ever since.'[135]

National and international correspondents started arriving at Kettering Grammar School by mid-morning on 21 December. Geoff and several Group members reenacted satellite tracking activities for the cameras and gave numerous interviews. The first BBC report with Geoff was for the 6:30 am news, and his last scheduled event that day was an Independent Television News (ITN) segment at 9:05 pm. The local Kettering newspaper headline at the top of the front page was 'Pentagon Signals A Hit!' with the associated story detailing the supposed American reaction to the Plesetsk discovery; however, the US government had no official comment about the announcement.[136] Geoff, Derek and several pupils were pictured in a banner photo accompanying the article.

Evening Telegraph

KETTERING'S SATELLITE TRACKERS

PENTAGON SIGNALS A HIT

Expert praises schoolboys

Kettering Evening Telegraph front page, Wednesday, 21 December 1966.

On the same day, the *New York Times* ran two articles about the launch site and the schoolboys' activities. One of the stories called the site Plesetsk, though there was no indication as to how the name had been determined; the reporter may have labelled it correctly with off-the-record help from the US intelligence community.[137] Geoff had not yet referred to the newly revealed base as Plesetsk—internally the Soviets were calling it that, as was the CIA—but he and his pupils would be forever linked to the no-longer-secret site. The Soviets did not publicly identify the Plesetsk launch site for another 17 years.

Plesetsk Revealed by the Soviets

ПО СТРАНЕ СОВЕТОВ

20 июня 1983 года

Место старта—Плесецк

ВЗГЛЯД СКВОЗЬ ГОДЫ

The article in *Pravda*, 20 June 1983, revealing the location and function of the Plesetsk launch site.

The Soviet Union finally acknowledged the existence of the Plesetsk site in two articles published in the state-run newspaper *Pravda* in June 1983. The first public mention by the Soviets of Plesetsk as an operational cosmodrome came in a feature article on the Kapustin Yar cosmodrome, published on 6 June 1983.[138] The article noted that 'In the early 1960s, yet another cosmodrome was established—Plesetsk—and it seemed that the 'space' history of Kapustin Yar came to an end'. The article went on to acknowledge the launching of satellites from Plesetsk, without providing further details.

Two weeks later a feature article entitled 'Place of Launch—Plesetsk' was published.[139] This article was written by the same author and appeared in the same 'Around the Soviet Country' section of *Pravda* as did the Kapustin Yar story. The 20 June 1983 article provided the first details published by the Soviet government of the satellite launching history of Plesetsk. The article opened with stories of pilots and fishermen seeing unusual phenomena in the night sky around Plesetsk, questioning if what they had seen was a 'flying saucer'.

The remainder of the narrative focused on the launch of the Intercomos-8 satellite nearly 11 years earlier, the difficulties encountered in the construction of the cosmodrome due to its northerly location and gave passing mention of the launching of Meteor weather satellites and Molniya communications satellites from Plesetsk. The article made no mention of the military purposes of the base, nor that the Kettering Group had informed the world of Plesetsk's existence many years before.

Until these articles were published in *Pravda*, Plesetsk had been a secret in the Soviet Union since it was first under construction starting in 1957. But the CIA had been aware of activity there for 25 years, and the Kettering Group had publicised its location and orbital launch function 17 years prior to this announcement. Why the Soviets did not admit to Plesetsk's existence earlier is an open question; perhaps it was a consequence of the large military presence there.

New Scientist magazine praised the Kettering Group's efforts. 'The fact that an enthusiastic group of boys, equipped with nothing more elaborate than an ex-War Department radio receiver, and a small amount of borrowed time on a computer, can make a real contribution of this kind…is a great vindication of the 'string and sealing wax' approach to research.'[140]

Geoff thought too much credit was being given to him and his pupils by all the press coverage about the reveal of the Plesetsk site. He wrote a note to the US Ambassador in London the next day apologizing for any inference that the US government had been unaware of the Plesetsk site or its capabilities. The Ambassador's science attaché wrote back, 'There is no cause to apologise for the embellishments which news accounts have added to your statements. On the contrary, I hope that the publicity given your achievements will serve as an example and incentive to others interested in pursuing space-related activities.'[141]

Sven Grahn and his schoolmate Bjorn Gimle tracked the latest Soviet launch, Cosmos 136, in December 1966. From their data, along with prediction bulletin information from the RSRS, Grahn and Gimle determined that Cosmos 136 had also been launched from Plesetsk. Their calculations further refined the location of the launch site, and their finding was published in *Flight International*—the first article by a Kettering Group member without Geoff as a co-author.[142]

Fast forward 22 years: Geoff and orbital analyst and Kettering Group member Max White in 1988 were working on figuring out from where Israel had launched its first satellite, Ofek 1; the Israelis had not disclosed the location. Geoff and Max had a somewhat easier time figuring out where the Israeli launch site was. White recalled:

> I went downstairs to the computer room and ran the orbit. I ripped it off and I went up to the library. In the library we had a copy of the *Times Atlas*. I got the atlas out and pulled one of the phones across and I rang Geoff. I said, 'I've got the *Times Atlas* here, [the launch site is] approximately thirty whatever north, whatever east.' The Americans were calling it the Negev Desert. He said, 'That's too broad, that's ridiculous.'...I could see there was this little railway siding that was going up by the beach. He said, 'You always look for a railway track. You've got to get your stages in somehow.' The siding went to a little place near the beach called Palmachim. I said, 'Let's call it Palmachim.' He said, 'We'll use that.'
>
> When we were at the Paris Air Show, these Israelis were there, and they've got various stuff and we're talking. We said, 'Where's your launch site?' 'Oh, we can't tell you that.' 'We know it's in the Negev Desert; we measured it out to Palmachim. He and I worked out where

it was.' We're thinking they're Mossad [Israeli intelligence agents] and they're thinking you're probably MI6. They just wouldn't discuss it. A couple of months later, we're at this meeting at Surrey with the amateur [radio] satellite guys, and there's a guy there from Israel, Israeli space agency, I think. We said, 'Where's your launch site?' He said, 'Oh, it's at Palmachim.' We start laughing and he asks why. 'Well, we found it but your guys at the Paris Air Show wouldn't tell us.'[143]

Location of Palmachim launch site in Israel.

6

Evolving Capability and Membership

1967 – 1994

In the pre-internet, pre-personal-computer era, manipulating numbers was much more time-consuming than using today's spreadsheets. Geoff was a believer, though, in using the tools available, emphasising comprehension over convenience. The pupils, we may assume, enjoyed the Kettering Group's technology upgrades made over time. Group membership changed over time also; mostly Kettering pupils passing through during their school years, along with a slow addition of UK-based and international enthusiasts who wanted to contribute to the Kettering Group's activities.

In the late 1960s, Geoff and Derek made several upgrades to the Kettering Group's shortwave equipment and the procedures used to obtain Doppler curves. They added a dedicated receiver for a time signal to improve the accuracy of their data, obtained a second HF receiver and modified the radios' antenna so it received signals near 20 MHz better. Geoff developed improved methods of using the Group's hardware to process signals. They also scrounged a vintage telegraphy signal recorder from the Post Office to aid in telemetry analysis—a Great Northern Telegraph Co. Ltd. Undulator Mk 2 (a 'pen recorder'), designed in about 1870 and, according to Geoff, manufactured in 1910. It was modified by Derek Slater to record telemetry from Soviet and Chinese satellites on a 10 mm-wide paper strip commonly known as ticker tape.[144]

Great Northern Telegraph Co. Ltd. Undulator Mk 2.

Derek's vintage Marconi CR100 shortwave receiver was becoming unreliable after many years of use, but the Group was getting by. Geoff described what happened to remedy the situation in the autumn of 1967:

> Old Cytringanian [alumnus of Kettering Grammar School] Robin Turner, of the Daily Express, happened to be in Kettering for his brother's wedding and found me at the labs on the Sunday, listening to signals. The CR100 was becoming temperamental, needing a smart tap with the hand from time to time, and Robin remarked that I 'could do with something better'. I told him that I knew of a second-hand RA17 [valve-powered radio from the 1950s] which could be had for the cost of refurbishment. I had already applied unsuccessfully for a grant to purchase it. Next day he phoned my wife and told her that he had persuaded his editor to 'go halves' with Racal [Electronics company] and give me a new receiver. I was lecturing to the BIS at the City University on the Tuesday evening. Before the lecture I was presented with a brand new, state of the art, transistorised RA217 by the chairman of Racal, (now Sir) Ernest Harrison and John Young, assistant to the editor of the Daily Express. Their gift removed all the uncertainty whilst retaining all of the pleasure.[145]

Geoff with Racal RA217 receiver, 1967.

The vintage CR100 radio was retired from satellite tracking activities, though it almost certainly continued in service with the Kettering Grammar School's radio club, which Derek supervised.

Robert Christy recalled a morning's tracking activities in an article for *Space Flight News* in the spring of 1968. He described getting to the physics laboratory before dawn and helping the Kettering Group listen for a re-entering Cosmos reconnaissance satellite that Christy had first detected a week prior. His article reflects the patience needed to find and interpret the faint signals from thousands of miles away. The authority figure in the story, 'Sir', is not named, but it is Geoff. He is depicted as kindly but exacting.

> Six minutes passed [since the last transmission from a descending satellite] before, over distant Kazakhstan, a parachute opened, pulling out with it another radio aerial. Back in Kettering, eager ears listened to the airwaves.
>
> Starting halfway through the letter 'K', a stream of Morse code came in. 'TK... TK... TK...' [TK was a standard transmission in these circumstances, as we shall see later.] An eight-foot diameter, heat-scarred sphere slowly dropped the last few thousand feet to Earth—'TK... TK... TK...' The signal varied in strength with a period of

around thirty seconds as the great ball rotated slowly under its single parachute—'TK... TK... TK...', 'Dah, dah-dee-dah, dah, dah-dee-dah, dah...' The 'T' wasn't real Morse code, its single dash was twice as long as it should have been, but then it always was, so who cared?

With a thump, the craft hit the ground. Simultaneously, the signal strength dropped as it rolled over and the aerial touched the Earth. Fabric billowed in a warm breeze, occasionally lifting the aerial wire off the ground, allowing the weak dots and dashes to get through.

'Let's see how long they take to find it', said 'Sir'. But on this occasion, it was not to be. Coming of day had brought with it unwanted noises from nearer home. Teleprinter code burbled away on a nearby frequency, getting stronger by the minute, and unintelligible voices flowed together and drowned out the plaintive 'come and find me' cry. In Kettering, it was all over.

Now it was 7:30 am. The boy mused to himself that 'Sir' was fine—he could have breakfast as he only lived half a mile from the school and had a car. For the boy, a trip home and back would be pointless. At least he could have another coffee! He stayed on, listening to the noise and hoping the beacon sounds would return. Occasionally he tuned to another Russian frequency. After all, they might have launched a Soyuz overnight, then he could be the first to hear it.[146]

In the early 1970s, Geoff and his pupils' activities came to the attention of the US Ambassador to the United Kingdom, Walter Annenberg. Annenberg hadn't wanted to be an ambassador until he was persuaded by President Nixon, and he had a rocky start to his tenure in London. He felt out of place, managed to anger the outgoing US ambassador and his wife, and put his foot in his mouth while speaking with the Queen. As a result, he wanted to quit and return to the US but was talked into staying on by US Secretary of State William Rogers.

After Annenberg paid out of pocket for a renovation of the US ambassador's mansion, England started warming to him. He became known as a philanthropist, and he looked for ways to contribute to the UK's well-being. Annenberg heard from the local press about the Kettering Group's lack of sophisticated tools. In early 1972, without any publicity, he gave Geoff the funds to buy a programmable electronic calculator, replacing their mechanical device.[147] The new Casio AL-2000 weighed almost 7 kilograms (15 pounds), sported 64 bytes of memory and could run a 30-step program. The new calculator reportedly cut computing time for a satellite's orbit from two and a half hours on the mechanical calculator to four minutes.

Geoff and pupil David Dean, centre, work with the Casio AL-2000 calculator, 1972.

Interesting titbits about the Kettering pupils and their mentor started appearing in the press, no doubt because of all the publicity the Kettering Group had received. An opinion/gossip column in the British journal *New Scientist* quoted Geoff as saying that his satellite tracking activities hadn't increased the number of pupils attracted to physics.[148] The columnist surmised that science teachers got so wrapped up in their pet projects that they didn't connect with their pupils. But a short blurb in *Aviation Week and Space Technology* the same month offered a counter to that conjecture: when a new Cosmos satellite launch was announced over a weekend, Geoff had motivated his pupils so that, in his absence, Robert Christy persuaded the Kettering school's caretaker to let him into the locked physics laboratory so he could monitor the spacecraft's signals.[149]

David Dean became a Kettering Group member in 1971, his second year at the school. Dean later recalled a unique opportunity that showed the trust Geoff had in him, and how Geoff was encouraging his pupils to share their findings:

> In late 1972 I was at home doing my homework and I heard someone ring the doorbell—my bedroom was directly above the front door—my father went to the door and I heard Geoff. This was 7 or 8 in the evening, and I wondered why he was coming round so late, what have I done wrong? I got on with my homework then Dad called me down. Geoff had come round to ask my parents' permission for me to appear on a TV programme called 'Magpie.' It was a programme for kids about cool things that were going on, current affairs and entertainment [focused on children]. They wanted to do a piece on the Satellite Tracking Group and my parents agreed.
>
> So a few days later, I went off to London along with Geoff's wife [Jean]. Geoff was at school, I took the day off, and his wife and I took the train with this enormous box containing a model Soyuz spacecraft which Geoff had made from detergent bottles and whatever. That was going to be our prime exhibit on this TV programme. We went to London, got to the studios. I was being chaperoned by his wife, and I did this little interview.
>
> My school friends from those days still remember this—sitting around the TV and there was David talking about satellite tracking in my school uniform.... That was one of the things where Geoff wanted us to gain confidence, speaking to the press, things like that which weren't naturally on the school curriculum. But they stood me in good stead in my subsequent life.[150]

In the early 1970s, Dean was pictured in an article about the Kettering Group published in *The Young Observer* section of *The Observer* newspaper. Geoff had quite a sense of humour, and he proceeded to pull the leg of the visiting *Young Observer* reporter. After demonstrating the Group's equipment and tracking techniques, Geoff and his pupils showed the reporter a photograph of the Moon's surface. The reporter was impressed and wrote, 'Prepared by now to believe anything, I asked Mr. Perry how he managed to take such a detailed shot. Giggles all around. It was a close-up of a piece of Kettering Grammar School chocolate pudding.'[151]

Geoff summarised the latest Kettering Group activities in a letter to *Spaceflight* in late 1972. He included highlights of the Group's activities since the 1950s. (The piece was apparently written in 1971 but held for later publishing, as he didn't mention the calculator received from the US ambassador several months earlier.) Additionally, Geoff noted that, up to October 1971, the Kettering Group had made 5,259 radio observations of 237 satellites. He described the mechanics of the Group's observations:

> The current basis of the work is the 90-minute routine monitoring session held every lunchtime during the school week for which the pupil members of the Group, in shifts, are responsible. Monitoring is also done outside school hours, at weekends and during the holidays. The work is extra-curricular and is carried out at school only because it is felt that much educational benefit is to be derived from having a successful research programme running in the science department.[152]

By the mid-1970s, Kettering Group members were making more independent contributions. Sven Grahn conducted an evaluation of the television transmission capability of Soviet spacecraft. From measurements of a spacecraft's television signal antennas, he deduced the frequency of that transmission. Sven described various purposes of the video signal and how the Soviets had used the video—monitoring animals during flight, watching two spaceships dock, and seeing how the cosmonauts fared during launch.[153]

Robert Christy began editing the 'Satellite Digest' section of *Spaceflight* magazine and would continue in this role through the late 1980s.[154] Since 1968, 'Satellite Digest' has been a thorough summary of all the world's recent space launches. Alongside his stewardship of the Satellite Digest in *Spaceflight* magazine, Christy started writing his own analyses of Soviet space programme operations. One of his first pieces was in the *Journal of the British Interplanetary Society* examining safety aspects of the Soyuz programme. He applied launch window constraints from the Apollo-Soyuz link-up several years earlier and developed data allowing predictions of future Soviet crewed mission launch times.[155]

Just before the Apollo-Soyuz Test Project (ASTP) mission in 1975 (we will learn later about Geoff's involvement with this project), Mark Severance, a 14-year-old US high school student from Fort Worth, Texas, contacted space historian and journalist James Oberg, who was organizing a satellite tracking group in the US. Mark was already listening to Chinese satellites on a multi-band radio and learning Russian in high school so he could understand what the cosmonauts were saying on the radio. He tracked the Apollo-Soyuz Test Project radio transmissions, along with the Salyut 5 space station and related Soyuz missions. James Oberg, who was running a satellite tracking network in the US, put Mark in touch with Sven Grahn and Richard Flagg. Later, at Geoff's invitation, Severance joined the Kettering Group.

Geoff managed to develop a contact at RAF Fylingdales, a missile tracking and warning facility in the UK, after a few visits there. Geoff would tell people cryptically about some information that he had received 'from up north.' His source would phone him occasionally and pass on information that he thought Geoff would appreciate. Apparently, the source's supervisor had a dim view of giving information to a bunch of schoolboys and discouraged the practice. But the link persisted through the 1970s and 1980s.

Another source of information for the Kettering Group was Pierre Neirinck, who worked at the RSRS in Slough assembling satellite prediction bulletins. Pierre also wrote a personal newsletter called Satellite Observing Notes, or SON. Desmond King-Hele said Neirinck was virtually a 24-hour service, which was appreciated by satellite trackers throughout Britain.[156] He had his own specialty: he followed low altitude US reconnaissance satellites that were difficult to track because of their frequent manoeuvres on orbit.

On 1 September 1976, Kettering Grammar School's name changed to Kettering Boy's School. The admissions process for the school had been selective, but it now was open to all male pupils. Geoff, recognizing that the Group's name, Kettering Grammar School Satellite Tracking Group, was no longer accurate, formally changed it to 'The Kettering Group'—although he had already been using that name in articles he had written.

The new name also indirectly recognised the international makeup of the Group, with representatives from Sweden, USA, Fiji, Germany and other locations. Geoff commented, 'At my suggestion, [fellow physics master] John Marshall produced the logo [for the Group] of a satellite in orbit around the rose from the Kettering Grammar School blazer badge. The centre of the rose was at the focus of the orbit since the Group was frequently the focus of media attention and the orbit was deliberately elliptical to support the popular belief that I am eccentric!'[157] The logo shows a satellite with a significantly higher apogee than perigee—the joke being that eccentricity is a property of a satellite's trajectory that describes the shape of its orbit. The logo featured on ties, other garments and sundries. Geoff also commissioned four silver brooches of the insignia.

Kettering Group insignia.

Jumpers with the Kettering Group insignia: Geoff, Jean, Charles, Isabel.

Geoff's sense of humour was given a solid test in late March 1977. A purported Russian television correspondent called the school and scheduled an interview with Geoff and his pupils. The story about the interview had run in the local Kettering newspaper, and everyone at the school was excited that the Russians were interested in the Kettering Group. But on the planned day of the interview, 31 March, no reporter showed up. School staff and pupils finally realised they had been pranked, as the next day was 1 April, April Fool's Day. The prankster, presumably someone at the school, was never identified. Letters arrived at the school on 1 April, taunting both the headmaster of the school and Geoff, who took the hoax in his stride. But he felt that the Russians may have been upset by the unflattering publicity, so he sent a letter of apology to the Soviet Union Ambassador in London.

The embassy wrote back, saying, 'I have been asked to acknowledge with thanks the receipt of your letter...and to accept your apologies. We do understand and appreciate popular traditions like "All Fools' Day." We only would like to hope that your letter, which incidentally is also dated 1 April, is not a hoax.'[158] Geoff's final words on the prank years later: 'To this day I do not know who perpetrated the hoax. I was not at all upset, to Jean's great surprise. In fact, it's just the sort of thing that I would have been proud to originate thirty years ago in my university student days during RAG Week [a student charity campaign involving pranks].'[159]

Group membership continued to evolve in the 1970s. Sarah Mobbs, daughter of friends of Geoff and Jean, wrote up a study in 1978 with Geoff about a collection of Cosmos satellites that flew at a 74-degree inclination and in a circular, 95-minute-period orbit.[160] Geoff and Sarah acknowledged Charles Sheldon's evaluation of this Cosmos class of missions as probably being electronic intelligence (ELINT) platforms—passive receivers of radio frequency signals. The US used a similar orbit to target Soviet and Chinese air defence radars. The Kettering authors analysed recent launches and put several more spacecraft in the ELINT category. Further information about the Soviets' ELINT platform was developed a year later by several other Kettering Group members. Newer Cosmos spacecraft with similar orbital parameters were identified as ELINT candidates and their active status verified by signal monitoring from Sven Grahn.

Mobbs had tinkered with radios when she was young, and Geoff had invited her to participate in pupil tracking sessions. She remembered how she came to write the paper with Geoff:

> I was always particularly fascinated why certain satellites were positioned in certain ways. In Geoff fashion he never used to tell you. He'd say, 'Why do you think these are spaced like this?' I said, 'These must be reconnaissance, these must be such and such.' So he said, 'I've got a good project for you. Why don't you look into these particular ones?' That's how it came about really. I used to sit there fascinated about the different spacings. I would sit and look at all the [elements of orbital information] that we used to get. Why certain ones were always replaced quite quickly if they ever went down. Geoff used to send me the two lines [orbital data parameters] and I used to go through them on the graphs. He'd say, 'What do you think is going on here, there were eight or ten originally and then they went down to four?' He said, 'I think you could write a paper on this.'[161]

Mobbs also remembered conducting visual tracking with Geoff. 'He was quite even[-handed], saying, "You do the radio tonight, you do the stopwatch tonight." He was very good at that. Particularly when we used to go down to Bude as well, because there was only Jean, me, my mother—who had no idea what we were doing at all—and my dad. People must have thought we were mad. We were all sitting on our back [outside] looking at where the [satellite] tracks were with the stop-watch, sometimes at 2 o'clock in the morning.'[162]

Geoff's last published piece prior to his retirement was a description of the pupil-run activities that the Kettering Group were engaged in that year. He updated all the subject areas of tracking and identified the pupils who were running specific projects during the 1983-84 school year.[163] Pupil Simon Turner had followed Cosmos communication satellite constellations during the year and detected both civilian and military networks. Group member Darren Conway monitored high-inclination spacecraft that were estimated to be electronic intelligence platforms. Sixth former Michael Beck was the last of four brothers to analyse a collection of high inclination, 12-hour period spacecraft that appeared to be similar in purpose to the US' Navstar Global Positioning System (GPS) navigation constellation.

Andrew Sims studied the high-eccentricity-orbit Molniya communication satellite fleet. Simon Parker analysed another set of high-eccentricity-orbit spacecraft and determined their mission was probably missile launch warning. Other areas studied by Kettering Group members included natural orbit decay, ocean reconnaissance satellite behaviour, cosmonaut voice transmissions and meteorology spacecraft. Geoff noted that Philip Perkins, with whom Geoff had published articles about Molniya communication satellites in the 1970s, had progressed to post-graduate studies at Cambridge, and that former Group member Andrew Driver was studying mechanical engineering.

Not many grammar school science masters have their retirement announced by *Aviation Week and Space Technology* magazine. From the 30 July 1984 edition:

> Soviet space monitoring operations at the Kettering Boys' School 70 mi. north of London have ceased following the retirement from the school of Geoffrey E. Perry, who for 27 of 30 years there led Kettering students in Soviet space tracking efforts. Those efforts pro-

> duced key discoveries on the Soviet space program including the location of the previously unidentified Plesetsk launch site.[164] [Note that pupil participation in tracking hadn't started until about eight years into Geoff's time at Kettering, rather than three years in this quote.]

Membership in the Kettering Group then became the exclusive realm of adult volunteers. A few years later, when asked if he had any regrets about shutting down pupil involvement in satellite tracking, Geoff said, 'Not really. I think that any teacher should have his own project and I'm sure my successors at the school will have their own ways of motivating children. The Kettering Group, however, hasn't closed. The adults [of the Group] around the world are still keeping an ear on the Americans and the Russians and the Chinese satellites.'[165]

The Kettering Group's membership was always informal, the Group consisting of whoever Geoff said was in it, and he was especially interested in getting international members on board. An avid satellite tracker in South Africa, Greg Roberts, started corresponding with Geoff in the late 1970s and became a valuable source of non-Soviet satellite information due to his location; the USSR's satellites tended not to transmit when over his part of the world. Roberts wrote about his involvement in the Kettering Group in 1986, after having given a talk to the Astronomical Society of South Africa on the topic of satellite observing. Years later, Roberts wrote about the highlights of his decades of satellite tracking, giving Geoff and the Kettering Group credit for spurring his interest and inviting him to contribute to the Group's projects.

In 1990, Geoff connected with British defence analyst Stuart Eves. Eves remembered his first contact with Geoff: 'One day [my manager] came into my office in London and said, "I've got this guy Geoff Perry on the phone. He understands a whole lot of stuff, and frankly I think you'll understand what he has to tell us much better than I would. Can I put him through to you?" "Yes, of course." [...] He had been listening to some transmissions associated with the Mir space station at the time. It took me about 30 seconds to realise that this guy knew a hell of a lot more about the Mir space station than I did.'[166] Eves was an active member of the Kettering Group for the next ten years.

Due to declining enrolment and competition with other local schools, the Kettering Boys School closed its doors permanently in 1993. The school's alumni association, the Old Cytringanians, compiled a history of the 400-year-old institution. Geoff wrote several pages of reminiscences, which featured the extra-curricular satellite research studies of the Kettering Group.[167] Also included were essays by Derek Slater, Sven Grahn, David Dean and Andrew Sims, recalling their involvement with Geoff and the Group. The book also provides an illuminating view of British schooling over the past few centuries. Dean and Sims had gone on to obtain doctorate degrees in scientific fields.

Geoff noted in his essay that with the advent of personal computers in the late 70s and early 80s, the Group had migrated their calculations to modern devices. As he described it, use of mainframe computers 'ended with the arrival of the Commodore 4032 which brought computing with programming in BASIC into the laboratory. Today [in 1995], members of the Group all have their own PCs with DOS or Windows. My daughter, remembering the excitement of the early days, dismisses the modern technological map displays and prediction printouts with, "I don't call that tracking!"'[168]

Part II

Satellites and Missions

From the late 1950s, when Geoff first sought to demonstrate the Doppler effect using a signal from a satellite, through 1966, when he and his pupils achieved a measure of fame, Geoff had taught his boys how collect and record information, analyse data and publicise results. Following the successful revelation to the world of the secret Soviet launch site, Geoff, Derek and the Kettering Group members were encouraged to further unravel the mysteries of the Soviet space programme. As time progressed, the lure of solving a large, continuous puzzle would drive the ever-expanding Group to increasingly sophisticated analyses. The following six chapters explore the Kettering Group's significant activities and detail their most notable findings; the dates in each chapter title show the years of activity of the Kettering Group related to that chapter's topic.

7

Early Telemetry Analysis

1962 – 1970

Early on, Geoff and Derek had been working on interpreting satellite telemetry behaviour—how and why the signals varied—though they couldn't yet understand the encoded information. Geoff recalled an early investigation effort, after Cosmos 5 passed over in 1962:

> As soon as I heard [Cosmos 5], it was obvious that its signals were of an unusual character. A periodic variation in pitch was superimposed on the normal Doppler shift. After his A levels, [Kettering pupil] Robert Jones made a study of the variation which indicated that it was due to a partly faulty solar battery. During a trip to Wales in the autumn half-term holiday, I called at the Physics Department at Aberystwyth [University], having read that they had also made observations of Cosmos 5. Ken Edwards, the chief technician, had a large collection of pen-recordings of the signals which showed [the telemetry] ceased completely when the satellite was in the Earth's shadow, confirming our hypothesis.[169]

Geoff believed that collecting and analysing data was pointless unless one's research was communicated, which he regularly demonstrated to his pupils. While tracking the Cosmos eight-day missions, Geoff and Group members noticed a peculiarity in the reception of signals during the recovery of the satellites. Through most of the year, the satellites' final transmission during the landing phase was audible at Kettering but during the winter months it was not.

Geoff and his pupils studied the data they had gathered and concluded that ionospheric conditions were driving this phenomenon. Since the Cosmos satellites were well below the local horizon during landing, their HF transmissions had to be reflected by the ionosphere to be received at Kettering. When this did not happen during the winter months, the only difference noted by the Group was the lack of sunlight illuminating the ionosphere between Kettering and the landing site in Kazakhstan, due to the Sun rising later in the winter at Kettering. It was known that the ionosphere's structure changes when sunlight was not energizing the layer above the location of a specific receiving site. Therefore, Geoff and his pupils concluded that the radio signals on the HF frequencies used by the satellites were not readily reflected when Kettering was in darkness, and thus less likely to be received. He wrote up this finding for the August 1966 edition of *Spaceflight*.[170]

Geoff's follow-up article in the December 1966 edition of *Spaceflight* showed how he and his colleagues were starting to analyse the information in satellite transmissions, as opposed to just observing the behaviour of the signals. With Derek Slater, John Marshall—a fellow Kettering Grammar School physics teacher—and Sven Grahn as co-authors, Geoff pondered the switching off and on of signals from several Cosmos satellites.[171] The authors presented data that suggested signal switching occurred on the last daylight pass over the Soviet Union and so may have been associated with enabling the reacquisition of the satellite's signal the next day. This paper foreshadowed more sophisticated analyses of Cosmos transmissions by the Kettering Group that would lead to numerous revelations about the Soviet space programme.

Geoff and Derek continued developing their satellite tracking methods for more accurate results. From the basic Doppler tracking method they described in their 1962 *Spaceflight* article, to the modifications described in a follow-up article in 1966, they wanted to continue to publicise their techniques. In a 1967 piece in *Wireless World*—the same magazine that published Arthur C. Clarke's 1945 geostationary communications satellite concept—Geoff described how the Soviets had modified their telemetry system.[172] Geoff and Derek had to determine the nature of the new telemetry signals and devise a better way of using the Doppler shift to calculate a satellite's ground track.

From a simple on-off switching of a satellite's radio signal, similar to tapping out Morse code, Soviet engineers had modified their telemetry mechanism from a one-frequency to a two-frequency system, initially making it more difficult for the Kettering operation to interpret the transmitted data and to use the Doppler shift to calculate the satellite's ground track. After analysis of the new signals, and understanding the principles involved, Geoff and Derek realised that telemetry values and Doppler shift could be still measured on a two-frequency system using a single radio receiver. (See Appendix B for more details on Soviet telemetry formats.)

Geoff and Sven Grahn co-authored a study in the *Spaceflight* December 1968 issue on how recoverable Cosmos satellites that were in space simultaneously either used different frequencies to distinguish their transmissions or were launched to ensure that only one satellite at a time was orbiting over the Soviet Union.[173] They noted that the timing of a frequency change allowed for the prediction of the impending launch of a new satellite. The authors mentioned that the satellites' frequency changes were so small that they would not have been detectable prior to the acquisition of the new shortwave radio from Racal Electronics and the *Daily Express* in the autumn of 1967.

The Kettering Group was becoming more intrigued in the late 1960s by the nature of the telemetry signals they had been listening to for several years. Geoff and the pupils had pen recordings by his contact Ken Edwards at Aberystwyth from their audio tapes of satellites. Geoff then acquired his own telemetry recording machine (the aforementioned Great Northern Telegraph Co. Ltd. Undulator Mk 2) to help analyse their collection of recorded transmissions, and a second shortwave radio to allow for multiple signals to be received.

Benefitting from new equipment and better analytical methods, Geoff wrote to *Flight International*, describing the telemetry of two Cosmos satellites tracked during their entry phase.[174] Cosmos 279's telemetry varied notably prior to landing; Geoff scrutinised the changes and mused on the meaning of those variations. He attempted to correlate a certain pattern in the telemetry with events in the satellite's re-entry process—the first time that the detailed structure of Soviet satellite signals had been analysed by the Kettering trackers.

The next mission, Cosmos 280, sent back standard Morse code characters in patterns that resembled earlier reconnaissance missions, but the increased complexity of the telemetry suggested a test of a second-generation spy platform. Geoff pondered whether the short time for which the Cosmos 280's recovery signals were sent indicated that the Soviets retrieved the satellite in mid-air rather than post-landing. However, Zenit-type spacecraft were too heavy for mid-air recovery—though later versions did use ejectable film canisters—and there is no evidence the Soviets ever used a recovery method similar to the American Discoverer satellite aerial snags.

Once, when Geoff and the Group members were trying to interpret a string of telemetry numbers, they were stumped. Stuart Eves recalled a story that Geoff had told him concerning how he had puzzled through the problem:

> To try and analyse [the data], they were printing it out on paper tape [using the Great Northern Telegraph Co. Ltd. Undulator Mk 2], getting these long strips of paper [of] Russian telemetry. He couldn't make much sense of one set of telemetry and didn't know what they were doing. He was sitting in the barber's having his hair cut, looking at his reflection and thinking about the fact that his reflection was reversed. Then he thought, ah, it's backwards! What they were doing was recording this telemetry onto a tape recorder on the spacecraft then running the tape backward through the transmitter in order to get it to the ground. What he had assumed was the least significant bit was the most significant bit, and vice versa.[175]

Kettering Group members David Muggleton, left, and David Dean, right, examine the Great Northern Telegraph Co. Ltd. Undulator Mk 2 output, c. 1973.

The Kettering Group continued their telemetry analyses through the 1970s and 1980s. Their studies revealed a hidden constellation of Soviet satellites, which we will learn about in Chapter 11.

8

Photo-Reconnaissance Spacecraft

1966 – 1998

EIGHT-DAY MISSIONS

For the first several years of the Zenit programme, the missions were mostly eight days in length, hence the 'eight-day wonder' label given to them by King-Hele and adopted by the Group. These missions were the bread and butter of the Kettering Group's activities in the 1960s. They were launched mid-day UK time; their telemetry signals were typically received during school hours and their landing transmissions helped pinpoint a satellite's end of mission.

Sven Grahn stumbled upon an unusual signal when monitoring Cosmos 114's last orbit in April 1966. The satellite's standard transmission had ended, but Sven left his receiver on 'not knowing what to expect,' as he admitted later. Several minutes passed, then he heard a Morse-code-like signal that lasted 15 minutes.[176] The code was equivalent to the letters T and K in both the Cyrillic and Roman alphabets. Listening for 'tee-kays' at the end of eight-day Cosmos/Zenit missions became routine.

Geoff and his pupils concluded that the signal was from Cosmos satellites during their return to Earth. Sometimes the tee-kays were during the last several minutes and sometimes for several hours; the Group determined that the signal was first heard when the spacecraft's parachute deployed during the entry sequence and the longer duration of the signals indicated recovery forces took more time to find the spacecraft.[177] Ground forces probably used the signal as a method of homing on the vehicle during its descent.

Sven Grahn recalled, 'You could hear when it landed because the signal strength went down when the parachute crumbled, because the antenna was in one of the parachute lines. When the antenna collapsed, the signal dropped. So we could even say the exact moment of touchdown.'[178] Why the letters T and K? In a 1968 *Spaceflight* article, Geoff and Grahn suggested that TK could be an abbreviation of the Russian phrase for 'ballistic flight completed.'[179] But, separately, Grahn also noted that other combinations of letters—TG and TF—were later transmitted by different versions of Zenit satellites. Grahn surmised that, since a colleague of his had a 1965 recording of the 'tee-kay' signals, the Soviets had probably been using the homing method since the first Zenit mission in 1962.

Geoff summarised his speculation about the purpose of the short-lived 'eight-day wonder' Cosmos flights in *Spaceflight* in mid-1968. Building on previous clues and bits of information, he laid out a convincing case for these missions being photo-reconnaissance in nature.[180] Geoff showed how ground tracks over the US overlap to achieve complete coverage of the country in eight days, and how the local time of passage over any given point coincided with relatively low Sun angles to provide shadows and contrast for photographs. He noted that, shortly after he presented a paper on this topic to the BIS in October 1967, Charles Sheldon's report to the US Congress on the Soviet space programme concurred that these satellites were in fact military observation platforms.

Separately, Geoff and Sven Grahn identified a telemetry value in a presumed Soviet reconnaissance satellite that decreased steadily during flight; they concluded this signal was a measure of an expendable onboard resource such as fuel or camera film. A few years later, Geoff wrote a section of Charles Sheldon's 1976 Congressional report on Soviet space activities, describing the telemetry characteristics of Soviet crewed space missions up to the mid-1970s.

12-DAY MISSIONS

Upgraded versions of the Zenit reconnaissance satellite commenced flying during 1968 with more sophisticated cameras and a longer mission duration. By 1970, almost all Zenit missions were about 12 days in length.

Figuring out what the ever-secretive Soviets were up to in space continued to be a popular activity in the West. Since Geoff and his pupils had been studying the Soviet reconnaissance satellite operations for several years, they had amassed a significant amount of data on the eight-day Cosmos missions and were beginning to study the new 12-day satellites. Geoff disagreed with a 1970 letter in *Aviation Week and Space Technology* by Saunders Kramer, a noted Soviet space programme analyst, that attempted to assess the latest behaviour of these platforms.[181] Kramer had grouped several 12-day satellite missions based on data announced by TASS. Geoff politely observed in his published reply that some of the letter writer's conclusions were 'questionable'.[182] He then summarised the Group's findings concerning the satellites' apparent flight characteristics and notable in-orbit manoeuvring. Geoff described satellite properties that were only apparent from direct satellite monitoring. Where Kramer relied upon publicised properties, Geoff noted that 'visual and radio observations of these satellites' revealed more information about their missions.

Continuing to monitor the short-duration Cosmos flights, Geoff summarised the Group's latest knowledge about the behaviour and purpose of these spacecraft. In the May 1972 issue of *Spaceflight*, he discussed data from the satellites' recovery beacons, telemetry interpretation, longer duration missions, in-orbit manoeuvring, changes in orbital inclination during the summer, and descent trajectories.[183] Geoff noted that most observation satellites were by then orbiting for 12 days, longer than the old 'eight-day wonders.' He theorised that longer duration was accompanied by a marked improvement in photographic resolution, since the ground tracks were more closely spaced and therefore a narrower field of view could be used by the cameras—resulting in photographs that were 'zoomed in' and showed finer details. These missions were in fact flown with improved cameras and better on-orbit manoeuvring capabilities; the new satellite was called Zenit 4M.[184]

Geoff further analysed the ground tracks of the presumed reconnaissance platforms Cosmos 463 and 464 which overflew south Asia during the Indo-Pakistani War of December 1971.[185] His research, based on Kettering Group-derived data as well as orbital information on the satellites from RAE and RSRS, indicated that both satellites had manoeuvred in orbit to make several passes over East Pakistan. Geoff suggested the reason for the manoeuvring would have been to have multiple looks at the ground situation during the war. He determined that film canisters would have been recovered within four hours of the last East Pakistan pass for each spacecraft.

Group member Stuart Eves recalled, 'The levels that [the Kettering Group] got to were quite sophisticated. They were able to identify which channels in telemetry were measuring particular things including on the surveillance satellites. They managed to work out which channels were the exposed film monitor, how much of the film was being used on the satellite. So they were able to predict when the re-entry capsule with the film was going to be ejected from the satellite because they knew how much film could go into each capsule.'[186]

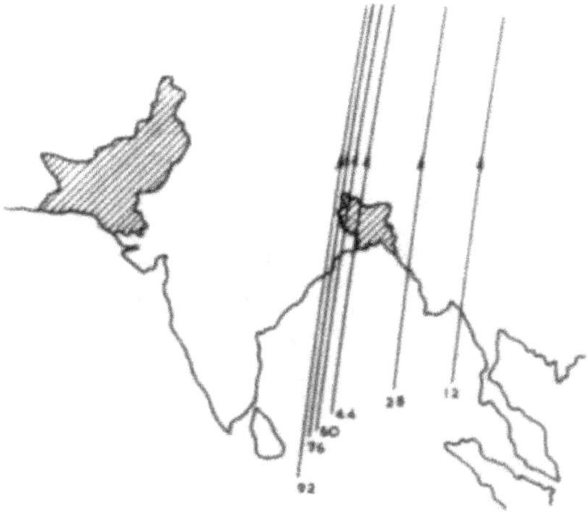

Cosmos 464 ground tracks over India and East Pakistan.

Geoff continued his detailed analysis of Cosmos satellite telemetry signals in 1974. He defined two classes of 12-day Cosmos missions, with and without manoeuvring capability.[187] Geoff found a word in the telemetry sequence that appeared to correspond to whether the satellite had an additional scientific mission. Such missions also appeared to release smaller fragments into orbit, at least one of which was presumably a scientific payload. *The RAE Table of Earth Satellites* shows numerous instrumented sub-satellites associated with Cosmos missions in the early 1970s.

Also in early 1974, Geoff analysed Cosmos spacecraft overflights of the Arab-Israeli conflict from the previous October.[188] He noted that several Cosmos satellites had manoeuvred over the battle area during the twenty-day-long conflict. Geoff and the Group listened to the satellites—although some were flying with no shortwave transmissions—and plotted their ground tracks with RSRS data after the war ended. Multiple Cosmos platforms had made numerous flyovers to look at the battlefields. Geoff described the orbit adjustments burns that allowed the spacecraft to photograph the conflict areas in detail.

ARKON-1

Geoff's Arkon-1 analysis is an example of how he updated and reused material. In 1998, *Aviation Week's* Craig Covault described a new Soviet reconnaissance satellite that had also captured Geoff's attention.[189] Called Arkon-1, it was as large as the Hubble Space Telescope but pointed at the Earth instead of deep space. Arkon-1 produced images with a five-metre (16.4 feet) to two-metre (6.5 feet) resolution from a medium altitude orbit. Covault credited research by the Kettering Group for helping to deduce the spacecraft's purpose.

With the assistance of Group members Stuart Eves and Grant Thomson, Geoff wrote an analysis of Arkon-1 for *The Imaging Science Journal of The Royal Photographic Society* in 1998.[190] He also followed this up with a longer piece in the *News Bulletin of the Astronautical Society of Western Australia*, 'And Whoever Heard About the Projects Zerkalo,

Nord and Arkon-1?'[191] (Zerkalo and Nord were proposed Soviet communication satellite networks.) Geoff went to the 1998 BIS Soviet/Chinese Technical Forum and talked about Arkon-1. Eves was also presenting at the 1998 Forum on 'The History of Molniya'; in his address, Stuart thanked Geoff for his contributions to the paper.

During the BIS Soviet/Chinese Technical Forum the following year, in Geoff's last known public talk, he updated his Arkon-1 presentation from the 1998 meeting. Geoff later published a revised version of the 'And Whoever Heard About the Projects Zerkalo, Nord and Arkon-1?' article in the April 1999 edition of the *Journal of the British Interplanetary Society.*[192]

9

Human Spaceflight

1967 – 1995

SOYUZ

In early 1967, Ken Edwards from Aberystwyth sent the Kettering Group pen recordings of an unusual signal he had recently heard. Geoff identified it as coming from Cosmos 140, which had been a rehearsal for the ill-fated April 1967 Soyuz 1 mission. Vladimir Komarov, who the Group heard talking on the radio during the earlier Voskhod 1 mission, perished during entry when Soyuz 1's parachute lines became tangled. When the Group received a pattern of signals in October 1967 from a new satellite, Cosmos 186, that was identical to that of Cosmos 140, Geoff concluded that Cosmos 186 was also a precursor to an upcoming human space mission.[193]

In a call with *Evening Standard* science correspondent Peter Fairley, Geoff mentioned what he had deduced about Cosmos 186; later that day, the *Evening Standard's* page one banner headline read 'Schoolboys Tune In To A Secret', the Soviets having not yet announced Cosmos 186's purpose. Fairley wrote, 'The first clue to suggest that it was not simply a Cosmos came from Kettering Grammar School, Northants, where physics master Mr. Geoffrey Perry and a team of boys picked up signals from it last night.'[194] A phone call from Geoff was enough to generate a front-page story, such was his credibility.

Other outlets, including *Aviation Week and Space Technology*, also credited the Kettering satellite watchers with scooping the Soviets for the second time in 12 months. Not only was Cosmos 186 a precursor to the Soviets' next crewed spaceflight, but it also participated in the first automated docking of two vehicles in space. Geoff and Sven Grahn monitored signals from both Cosmos 186 and the vehicle it docked with, Cosmos 188.[195] The next crewed Soyuz mission did not occur for another 12 months, the Soviets evidently having significant hardware problems to correct after the Soyuz 1 tragedy. When Cosmos 186 was launched, both *The Times* and the *New York Times* quoted Geoff commenting on possible mission goals and pupil tracking operations. Geoff and the schoolboys also made the front page of the *Evening Standard* again, for their later revelation of further Cosmos 186 activities.[196]

Geoff and the Group continued developing more sophisticated analyses of Soviet satellite telemetry. In late 1970, Geoff and overseas Kettering Group member Richard Flagg published a technical article about interpretation of telemetry from several different Soviet spacecraft. Flagg was a radio astronomer and systems engineer at the University of Florida. He had access to sophisticated state-of-the-art receivers, instrumentation and strip chart recorders. He also had ample experience with satellite tracking and telemetry data reduction. Using both the vintage and state-of-the-art strip chart recorders to which they individually had access, Geoff and Flagg could see varying pulses in the signals and looked for patterns in chart recordings.

Telemetry from Soyuz spacecraft was studied in detail.[197] They determined the Soyuz 3 mission used variable length pulses and there were 32 possible values for a given pulse, with the length of each pulse representing the value of a specific parameter measurement from onboard the spacecraft. They also found telemetry indicating when cosmonauts had transferred from the Soyuz 5 spacecraft to Soyuz 4 after the two spacecraft had docked. The two researchers found evidence for occasional transmission of cosmonaut biomedical data, likely heart or respiration rates. The telemetry also provided an identifier for which cosmonaut the biomedical data pertained to on flights with 2 or 3 person crews.

Geoff was always looking for new patterns in the data collected by the Kettering satellite trackers. He noted in a December 1970 article in

Flight International that most Cosmos missions were recognisable by their orbital characteristics, and those that weren't recognizable were typically failed missions or tests of new capabilities. By combining known information with the Group's radio and visual observations, Geoff analysed recent outlier Cosmos spacecraft that didn't fit known categories. There had been rumours of the Soviets intending to send a space station into orbit, and Western observers were anticipating indications of a station in certain Cosmos satellites' behaviour. However, Geoff came to no conclusions about unusual spacecraft properties and manoeuvres, noting simply 'further developments are awaited with interest.'[198] The Soviets had in fact started working on a crewed space station, but it was several months away from being launched.

As well as keeping close tabs on the USSR's robotic missions, Geoff and his pupils were continually following the Soviet human spaceflight programme. Since the Soyuz spacecraft transmitted the cosmonauts' voices on shortwave frequencies similar to those that Cosmos satellite telemetry used (in addition to very high frequencies (VHF)), the Kettering trackers could tune in, especially when the cosmonauts flew over the UK. Geoff and the schoolboys undoubtedly followed the events as the Soviet Union launched the world's first orbiting space station, Salyut 1, in 1971. (More on the Salyut stations in the next section.) Four days later, Soyuz 10, with three cosmonauts aboard, attempted to dock with the station but, though the vehicles were attached to one another, the docking process couldn't be completed. Mission controllers had to devise a work-around method to successfully undock, and the crew returned to Earth.[199]

Several weeks later, after the technical difficulties had been addressed, Soyuz 11 lifted off to dock with Salyut 1. The Kettering trackers had anticipated the launch by calculating when Salyut 1 would pass over the Baikonur launch complex and were monitoring the typical Soyuz voice radio frequencies. Robert Christy, who had reached the end of his time at school but was still involved in tracking operations, heard the first signal from Soyuz 11 about ten minutes after launch; the high frequency Soyuz telemetry was recognizable, the Group having monitored the same telemetry format since the Cosmos 140 launch.

The next morning, the front page of *The Times* proclaimed that Geoff had announced the launch of Soyuz 11 an hour before the Soviets did.[200] A few days later, *The Times Educational Supplement* featured a photo of Robert Christy intently listening to Soyuz transmissions on the Kettering Group's HF receiver. The photo's caption read, 'Jodrell Bank [the famed British radio telescope facility] congratulated the Kettering team for being the first in the West to pick up the [Soyuz 11] signals.'[201]

Jodrell Bank Observatory

The UK's Jodrell Bank Observatory was, and still is, one of the world's leading centres for radio astronomy. In the 1960s, the observatory operated the largest existing steerable antenna, the 76-metre (250 feet) Mark I radio telescope, renamed later as the Lovell Telescope after Sir Bernard Lovell, the founder of the Jodrell Bank. The facility regularly monitored early Soviet satellites.

The Lovell Telescope at Jodrell Bank Observatory.

The Soyuz 11 crew made history as the first humans to occupy and operate a space station. They worked aboard Salyut 1 for three weeks, then prepared to return to the landing zone in Kazakhstan. The mission appeared to have gone well, although both Sven Grahn and space historian Phillip Clark reported there were indications of a fire on the station that the Soviets never mentioned.[202] The Soyuz 11 vehicle departed Salyut 1 having completed an array of scientific and medical investigations. Shortly after their deorbit rocket fired, the descent module separated from the rest of the spacecraft. No radio transmissions from the crew were heard during the descent and landing, worrying officials on the ground.

When recovery personnel reached the spacecraft, they found the Soyuz 11 crew dead. Geoff and several pupils had listened to Soyuz 11 telemetry on its last orbit, just prior to module separation. In *The Times* report on the tragedy, Geoff was quoted as saying, 'At that time, we are certain that all three of the men were still living.'[203] The investigation into the Soyuz 11 deaths showed that a cabin vent valve had opened prematurely as the vehicles separated and the crew had been asphyxiated. Later, Geoff said, 'We felt we'd lost friends. Although we never met them personally, we'd listened to their telemetry for 23 days. It was a very sad, sad event indeed.'[204]

Geoff and his pupils made the front page of *The Times* newspaper yet again when the Soviets sent Soyuz 14 aloft in 1974 with two cosmonauts on board.[205] Soyuz 14's goal was to dock with the newly orbited Salyut 3 space station, a military observation platform (part of a classified programme called Almaz) in the guise of the Soviets' civilian programme. The Kettering Group had predicted likely Soyuz launch windows for station rendezvous and were monitoring their radios when the cosmonauts departed. While Soviet officials would not specify the launch time, Geoff told *The Times* the precise moment of liftoff for Soyuz 14 based on the Group's radio monitoring. Geoff noted that the telemetry and voice signals from the spacecraft were consistent with earlier Soyuz crewed missions. The Soyuz 14 crew spent two weeks on board Salyut 3.

The next crew bound for Salyut 3 had a more difficult time. The

Soyuz 15 mission launched with the presumed intention of spending two to three weeks aboard Salyut 3, a normal mission duration. In Sweden, Sven Grahn heard the cosmonauts talking to their mission control shortly after launch. Geoff awoke early the next morning and saw in the predawn sky two orbiting vehicles passing overhead—Soyuz 15 had apparently failed to dock with the station.[206] The next day, the Soyuz crew returned to Earth, making the first ever Soviet human mission night landing. Soviet space officials claimed the crew was never supposed to dock with Salyut, only evaluate new on-board systems, and that the night landing had also been intentional.

Geoff determined that the launch time and pattern of Soyuz missions to Salyut stations had pointed toward a docking and a several-week mission. Soviet officials concealed the truth about the mission failure for many years: the automatic rendezvous system on board Salyut 3 failed, resulting in Soyuz 15 approaching the space station at a dangerous speed multiple times. Salyut 3 remained unoccupied and was deorbited over the Pacific Ocean in January 1975. Neither of the Soyuz 15 crew members flew in space again.[207]

An unknown aspect of Soviet space operations was revealed by the Kettering Group in 1991. Several years before, Sven Grahn and Mark Severance had heard radio transmissions from a Soyuz crew that implied they were to land in the US in case of emergency. Later, Sven asked a visiting cosmonaut about a term he'd heard on the radio, 'ugol pasadki', which means 'landing angle' in Russian and relates to the exact location of a spacecraft's touchdown. The cosmonaut confirmed what Grahn and Severance had suspected: in an emergency, a Soyuz spacecraft could have landed in a Nebraska corn field.[208] Soviet space planners were targeting the US' upper Midwest as an emergency landing area for orbits where the crews did not overfly the USSR. Further research by Geoff and Group members showed that many Soyuz missions had a daily plan for emergency descent not just into the US and Canada, but also potentially in western Europe and the seas around Japan.

Nothing of this plan had previously been known to open sources, and *Aviation Week* noted that NASA had never discussed the US landing option with Soviet officials.[209] However, as early as the mid-1970s the Soviets had prepared for the possibility of a Soyuz landing and crew recovery outside of the nominal landing zones in the Soviet Union. Joint documentation from the Apollo-Soyuz Test Project illustrated the placement of Russian- and English-language placards on the bottom of Soyuz descent module to aid in the extraction of the crew after landing. In addition to these placards, which were exposed after jettison of the spacecraft's heatshield during descent, special tools were attached to enable the opening of the hatch at the top of the descent module to facilitate crew egress.

Prior to the start of crew operations on the International Space Station (ISS), joint coordination procedures were developed by the NASA and Russian flight control teams to facilitate crew recoveries in the event of a contingency landing in the US or other territories outside of the normal Soyuz landing zones in Kazakhstan. Mark Severance led the initial development of these procedures with his Russian counterparts.

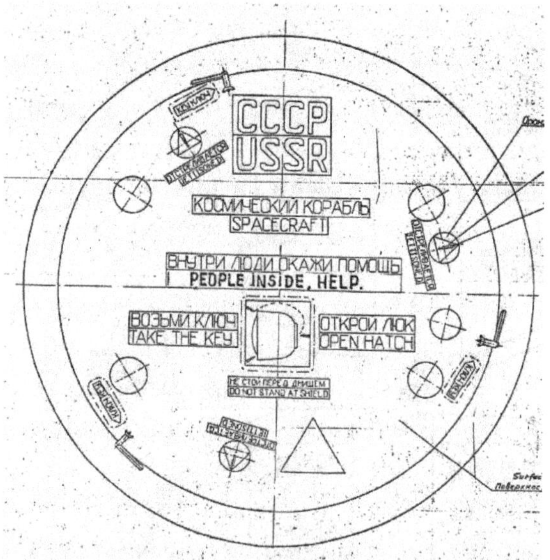

This placard was intended to assist the rescue of cosmonauts if they landed outside of the normal Soyuz landing zone in the Soviet Union.

SALYUT

Salyut space stations

The Salyut series of space stations had hundreds of instruments and experiments on board each vehicle—scientific, medical and biological. Salyut 1 included the first crew-operated telescope outside the Earth's atmosphere. Salyut 2 failed in orbit, apparently due to explosion of its booster. Salyuts 3 and 5, known internally as Almaz along with Salyut 2, had combined civilian and military instruments and multiple reconnaissance cameras, and operated at a lower altitude for better photography.

Salyut 3 included an automatic machine gun mounted at the front of the station, in keeping with its military mission, but it proved impractical. Each successive Salyut contained increasingly sophisticated equipment and allowed longer crew stays. Salyuts 4 and 6, the 'civilian variant' of the Salyut stations, served to incrementally extend the durations of crew missions in period of 1974–1982.

Salyut 6 introduced the regular occurrence of extra vehicular activities (EVAs) for maintenance and external payload installations, as well as regular resupply flights from the uncrewed Progress cargo vehicles. Additionally, Salyut 6 and 7 demonstrated the addition of modules to the core station. The addition of these modules, which were comparable in size to the Salyut stations themselves, paved the way for the modular assembly of the Mir space station and ultimately the Russian Segment of the ISS. The last Salyut vehicle, Salyut 7, hosted the longest duration crew of the Salyut programme (237 days) in 1984 and the final crew of the Salyut programme in 1986. Salyut 7 deorbited over South America in 1991.

Salyut 7 space station.

The earliest alumni of the Kettering Group were starting their careers by the 1970s. Sven Grahn had graduated from the Royal Institute of Technology in Stockholm with a master's degree in engineering physics. In 1974, he wrote about the future of the Salyut space station series. Grahn discussed characteristics of, and changes in, the Salyut station configuration through Salyut 3.[210] There was no notion at that time of the military nature of some of the Salyut missions, but Grahn thoroughly analysed available information to paint a clear picture of what the Soviets were probably doing in their space station project. Within a year, Sven would join the Swedish Space Corporation as a spacecraft engineer.

Keeping an ever-keen eye on the Cosmos satellite series, an article by-lined 'The Kettering Group'—the first time in print that this name was used—in the January 1974 issue of *Spaceflight* described an unusual set of characteristics for the May 1973 mission of Cosmos 557.[211] It was transmitting on frequencies that were typically used for human spaceflight, and the Group showed that the telemetry type and values coincided with both Soyuz missions and the Salyut 1 space station. But no crew was aboard Cosmos 557 and the vehicle decayed out of orbit after several days. Geoff and his pupils learned that Cosmos 557 had been launched on the Soviets' largest booster, called Proton, so they concluded the satellite was a Salyut-type space station that had malfunctioned and proved unusable. In fact, it was supposed to be called Salyut 3; the name change to Cosmos 557 was to conceal the failure.

Regarding the 'Kettering Group' by-line, Max White noted later, 'Geoff felt that he might write up the article, but with all the input from people, like data analysis, division, etc., etc., he would just put [it] under the generic 'The Kettering Group.' He felt that everybody was part of it and he didn't exclude people. So the big ethos was information sharing; if you had a nugget, you would always pass it on.'[212]

When Richard Flagg and his family passed through Sweden on holiday in 1978, they visited Sven Grahn. Flagg and Grahn had heard cosmonaut transmissions from Salyut 6 that were unintelligible and sounded like 'Donald Duck talking with a sore throat.' They determined that the most likely voice scrambling technique was inverting the audio spectrum—electronically changing high frequencies to low and vice versa. Flagg remembered:

> Late one night we went into the electronics lab at Sven's workplace with a tape recording of what we believed was inverted speech from Salyut 6. After a few hours of work in the well-equipped lab we fabricated a device to un-invert the voice signals. It worked—Donald Duck was gone—replaced by an intelligible voice. Our suspicion was right as to the method the Russians were using to scramble the voice signal. It was a simple yet effective way for them to assure privacy in conversations between the cosmonauts and doctors at Russian mission control.[213]

Two years later, pupil Andrew Driver was assigned to analyse the behaviour of Salyut 6, the first of the Soviet's second-generation space stations. He and Geoff compiled their observations and analysis of Salyut 6 manoeuvres in the *Journal of the British Interplanetary Society.* [214] They studied orbital data provided by NASA, information from Radio Moscow announcements, and a colleague's calculations of Salyut 6's orbit. The satellite trackers noted certain changes in Salyut's orbit occurred just before the launch of crewed missions, and prior to departures returning a crew to Earth. They also determined that Salyut 6 typically had a 31-orbit repeating ground track, where the station crossed over the same swath of the planet every other day. According to British space expert Phillip Clark, 'One of the major experiments in the Salyut programme has been the observation of the Earth (whether for military or scientific/economic reasons) and therefore the orbits of the Salyuts are chosen to allow the stations to fly over the same areas at specific intervals.'[215] Geoff and Driver continued their observations of Salyut 6 and published a further analysis of its orbit in 1981.

In 1982, the Soviets replaced their Salyut 6 space station with Salyut 7. Salyut 7 hosted several crews, including one that set a human spaceflight record for the longest mission at that time. *Aviation Week* noted that an addition to the space station had been launched and quoted Geoff describing the new section's telemetry data. The fourth crewed mission to Salyut 7 failed to dock with the station; Geoff commented in the *New York Times* that 'presumably something went wrong with the rendezvous and reaction control system.'[216] He also noted that, while monitoring the Soyuz crews' voice communications, they 'weren't their usual normal jolly selves.' *Aviation Week* cited Geoff's observations describing the crew's orbital manoeuvres during the attempted rendezvous.[217] Phillip Clark later wrote that the Soyuz's rendezvous radar had malfunctioned due to a damaged antenna rendering the automatic system inoperative. When the crew missed their first manual docking attempt, they were too low on fuel to make a second try and proceeded to successfully deorbit and land.[218]

Even after Geoff's retirement from teaching in 1984, he remained in demand as an expert consultant for journalists around the world. In May 1986, *Aviation Week and Space Technology* published an article about how Soviet cosmonauts flew their Soyuz T-15 spacecraft from the new Mir space station to the older Salyut 7 vehicle. Geoff is quoted throughout the article, describing observations he had made of the transfer and discussing possible reasons, including finishing uncompleted tasks on Salyut and readying a module attached to Salyut for a move to Mir. He called the cosmonauts' activities 'a great leap ahead in manned space' experience.[219]

MIR

Geoff was again the expert of choice for confirmation of cosmonaut activity related to the new Mir space station. The *Aerospace Daily* journal reported that two cosmonauts had boarded Mir, according to an announcement from TASS. Geoff described the location where the crew docked with Mir—over China rather than, as was typical, near the Crimea communication centre—and noted that the crew stayed in their spacecraft for two orbits before moving into Mir.[220] The article also cited Kettering member Dave Hawkins reporting activity on Mir's uplink frequency used for voice and teleprinter functions, further confirming occupation of the station.

Mir

The Mir space station was larger and more complex than its Salyut predecessors. Over its fifteen-year lifetime, 1986-2001, Mir hosted 28 Soviet and Russian long-duration missions and numerous other Asian, European and North American crews. The United States flew several Shuttle missions to Mir (Russian for 'peace' and 'world'), dropping off and picking up US astronauts who spent several months on the station. This cooperation between the two countries set the stage for the joint US-Russian construction and operation of the International Space Station. Kettering Group member Mark Severance participated in the NASA-Mir programme and designed the programme's patch.

Mir space station.

The mission patch for the NASA-Mir programme was designed by Mark Severance.

Geoff continued his expert commentary on Soviet crewed spacecraft activities. With two cosmonauts on board, the Mir space station had a new laboratory ferried up to it. The Kvant module, dedicated to astrophysics observations and materials science experiments, was only able to complete a partial dock with Mir. Four days later, on the second try, the module was successfully connected with Mir. Geoff listened to the cosmonauts' voice transmissions and provided *Aviation Week* with an analysis of the crew's activities and mood. He pronounced the cosmonauts to be 'in good spirits following the successful docking.'[221]

Geoff and his Kettering colleagues continued their monitoring activities, especially using VHF radio receivers to listen to cosmonaut transmissions from Soviet spacecraft, including Mir. In mid-1995, two cosmonauts were ferried up to Mir on the US Space Shuttle *Atlantis.* This was the first time the Space Shuttle had docked with Mir. During a spacewalk to deploy a scientific instrument, one of the two cosmonauts had a space suit malfunction, and the spacewalk ended prematurely. Geoff—and presumably other Kettering Group members around the world—listened to the cosmonauts' voice communications and confirmed for *Aviation Week* that the spacewalk had been curtailed, and the goals of the spacewalk had not been fully achieved.[222]

10

Individual Mission Analyses

1968 – 1995

FRACTIONAL ORBITAL BOMBARDMENT SYSTEM

A 1968 report in *Aviation Week and Space Technology* had identified ICBM defence and the 'three quarters ICBM' mission—nuclear-tipped missiles flying to the US over the South Pole rather than the North Pole—as possible purposes for a series of single orbit Cosmos flights.[223] In a letter to *Aviation Week* commenting on the article, Geoff suggested that coordination with military manoeuvres for missile defence might explain some of the missions, and that testing of the 'three quarters' profile may have continued with three more recent one-orbit Cosmos launches.[224]

Subsequently, the US Defense Department described the latest of those three missions, Cosmos 183, as being part of the Soviets' evaluation of a Fractional Orbital Bombardment System (FOBS)—the three quarters ICBM trajectory. FOBS was a method of delivering weapons from Earth orbit, making their detection more difficult. 'Fractional' implied less than one full orbit, a trajectory that was necessary to avoid infringing on the newly enacted Outer Space Treaty. For strategic and technical reasons, the Soviets abandoned FOBS in the early 1980s.

CHINESE SATELLITES

Continuing his analyses of satellite telemetry, Geoff described in the journal *Nature*, one of the most influential science journals in the world, signals from China's first satellite, launched in 1970.[225] Along with an electronic rendition of China's national anthem, 'The East is Red', ten seconds of telemetry from the spacecraft was transmitted every minute. The Kettering satellite trackers listened to China's new satellite during its 12th orbit through its end of life. With the Kettering Group's increasing sophistication in data recording and reduction, Geoff and his pupils noted certain details of the satellite's telemetry signal. The radio pulses appeared to have 32 values, implying the telemetry was formatted in 5-bit data streams (see below).

At least one report suggested that Geoff and his pupils had 'scooped' the Chinese by announcing the flight publicly; however, the Chinese government had issued a press release a few hours after the launch that was picked up by international media. The Kettering trackers started listening to the spacecraft 90 minutes after the announcement, having been tipped off to the launch and provided the satellite's transmission frequency by a *Kettering Evening Telegraph* correspondent. [226]

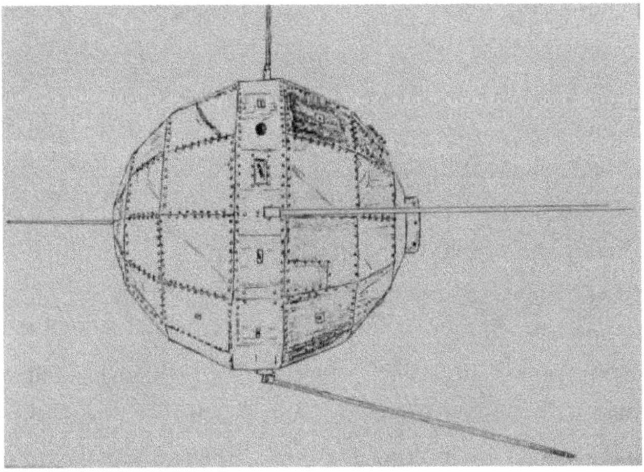

China 1 satellite, drawing by Geoff Perry.

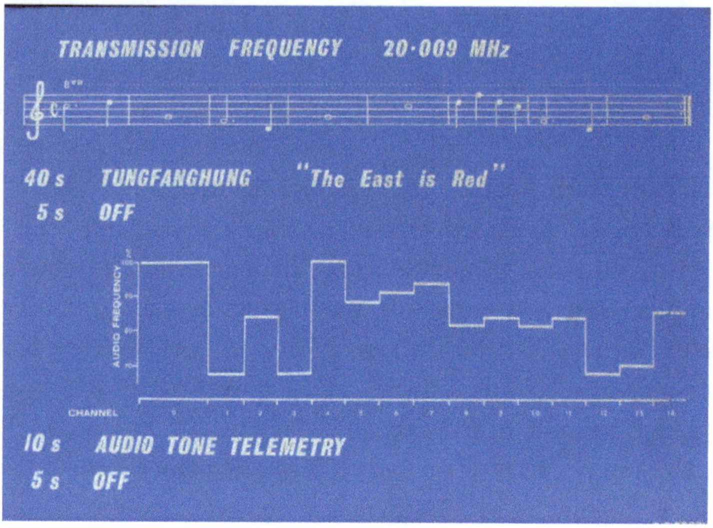

Telemetry from the China 1 satellite for 'The East Is Red.'

5-bit Data

We normally count in base 10, using the digits 0 to 9. Electronic devices like satellites and computers use binary, or base 2, counting, because only two digits are needed: 0 and 1. Binary digits can be represented by turning a transmitter on (representing the value 1) or turning it off (representing the value 0), or by alternating transmissions between two closely spaced frequencies: the higher frequency on and lower frequency off (1), or the higher frequency off and lower frequency on (0). A 5-bit binary system, having five bits per value, would start with: 00000 (0), 00001 (1), 00010 (2), etc. up to 11111 (31), for a total of 32 values. (By extension, 6-bit data allows for 64 values, 7-bit data for 128, and so on.)

Geoff and his pupils recorded and transcribed Chinese satellite telemetry, as well as Cosmos spacecraft transmissions, and tried to deduce some of the meaning of the signals by looking for binary-related patterns in pen recordings. This was the heart of the Kettering Group's telemetry reduction method.

Geoff and Richard Flagg published several articles in *Space Communications and Broadcasting* in July 1987. Geoff wrote a summary of Group activities and a new analysis of the second Chinese satellite that had been launched in 1971.[227] Flagg described in detail telemetry from the first two Chinese spacecraft.[228]

Geoff commented for *Aviation Week* in August 1992 about a new Chinese spacecraft with a separate recoverable module, estimating the new module would return to Earth in 15 days.[229] Geoff was again a resource for *Aviation Week* in 1993 when the Chinese reconnaissance satellite FSW-1 malfunctioned during a re-entry burn and later descended uncontrolled into the atmosphere. He predicted that two tons of debris from the satellite would land intact after telemetry received by the Kettering Group showed that the spacecraft had not separated as planned and was tumbling out of control.[230] Later analysis showed that substantial portions of the satellite had impacted the Pacific Ocean near Peru.[231]

PROSPERO

Geoff's daughter Isabel was attending Kettering High School in the early 1970s. Over several months, Isabel, along with Geoff and Derek Slater, monitored the signals from Britain's first satellite, Prospero, and followed the increase in its period of rotation; Prospero spun about its central axis to help maintain the vehicle's attitude. The spacecraft had lost one of its four antennas during ascent, so its radio signal's strength varied as the satellite rotated. Geoff later recalled how Isabel came to lead the Prospero investigation:

> In 1971, the British launched their one and only satellite on their one and only rocket from the Woomera rocket range [in Australia]. We thought that we better listen to this. We went to the school, pointed the six-element Yagi antenna in the direction the satellite would come, turned on the Eddystone 770R [receiver], picked up a signal, recorded it, turned the receiver off and went home. This was

> a Saturday. In the middle of that afternoon, I said to my daughter, 'That was spinning. That was spin rate we were listening to.' She said, 'Aren't you going to tell Uncle Derek?' I said, 'Yeah, I'll tell him.'
>
> I walked down the hall to the phone and as I put my hand to the phone it rang before I could pick it up. It was Derek, and he said, 'That was spin rate.' We had come to same conclusion almost simultaneously, practically telepathically. So [I] got in the car, up to the school, turned on the pen recorder, found it was spinning at 182 rpm.... This was pretty close, the nominal rate for Prospero was 200 rpm at launch. We checked with... [the] RAE on Monday and in fact the spin rate was a little lower than it should be.
>
> The reason why we were getting this throbbing was because... part of the Black Arrow [final stage] performed better than it should, it didn't shut down, overtook the satellite and knocked one of the four antennae off.... I said to my daughter, 'This would make a good project for one of the boys because, due to the Earth's magnetic field, eddy currents will be induced in the shell of the satellite, which will cause it to slow down. If we check this for a year, we'll be able to see the satellite spin [rate] decay exponentially.' She said, 'Why can't I do that?' I said 'Well, you can if you like', so at the age of 16 she started her project on the spin rate of the Prospero satellite.[232]

Isabel recorded the Prospero signals regularly for nearly a year, helped in the data reduction and analysis, and was a co-author of the paper 'Measurements of Spin Rate of the Prospero Satellite' published in the *Journal of the British Interplanetary Society,* along with Geoff and Derek. Displaying a bit of national pride, the first sentence of their article reads, 'Motivated by patriotism rather than an expectation of being able to resolve telemetry, an attempt was made to receive the 137.56 MHz transmission from 1971-93A, Prospero, on the day of launch.'[233]

Prospero

Prospero was a British satellite conceived in the 1960s to test advanced communication satellite technologies. Launched on a British-designed and -built Black Arrow booster from Australia's Woomera rocket range in 1971, the satellite was operational for two years until its tape recorders failed. The spacecraft was still transmitting on its main frequency when the end of the mission was declared in 1996. Prospero was built at the RAE Farnborough's Space Department, where Desmond King-Hele worked, and he was involved with the analysis of Prospero's orbit. Ironically, the Black Arrow rocket project had already been cancelled when Prospero was launched and was the only orbital-class rocket ever built and flown by Britain to date.[234]

Prospero satellite flight spare in the Science Museum, London.

COSMOS 482

In an early 1972 issue of *Flight International,* Geoff wrote a short analysis of the Soviets' star-crossed attempts to send a probe to Venus.[235] He compared the Venera 8 satellite to Cosmos 482, which had launched four days later. Geoff concluded that both had been intended to fly to Venus, as the Soviets typically sent two interplanetary spacecraft per launch window for redundancy. They often renamed a planetary mission if it failed, which appeared to have been the case for Cosmos 482. Venera 8 did indeed fly successfully to Venus and was the world's second spacecraft to transmit data from the surface of another planet, the first being Venera 7.

Geoff analysed the known data for Cosmos 482 and concluded that the satellite had started on its interplanetary trajectory but failed to gain enough velocity to leave Earth's orbit. A similar conclusion was published almost simultaneously in *Aviation Week and Space Technology,* validating Geoff's finding.[236] Cosmos 482 continued circling the Earth until 10 May 2025, when the spacecraft made an uncontrolled entry into the atmosphere. A Russian announcement said that the vehicle had descended into the east Indian Ocean near Indonesia.[237]

Barely 48 hours after the launch of Cosmos 482, several pieces of space debris landed near Ashburton, New Zealand. The local farmers who found the objects suspected they were either contaminated or radioactive and called in authorities to handle the wreckage. Since the objects were spherical, residents and local media dubbed them 'the Ashburton balls'. Experts at New Zealand's Department of Scientific and Industrial Research (DSIR) determined that the balls were composed primarily of titanium and probably came from rockets launched by either the Soviets or the US.[238] However, both countries denied any connection to the errant spheres. Geoff was contacted by a private individual in New Zealand who asked him to determine the source of the objects.

After Geoff gathered sufficient information from news reports and a summary of findings from the DSIR, he concluded that the only possible mission that could have produced the debris was the launch of Cosmos 482.[239] He based his conclusion on the timing and trajectory of the Cosmos 482 spacecraft and its rocket body parts.

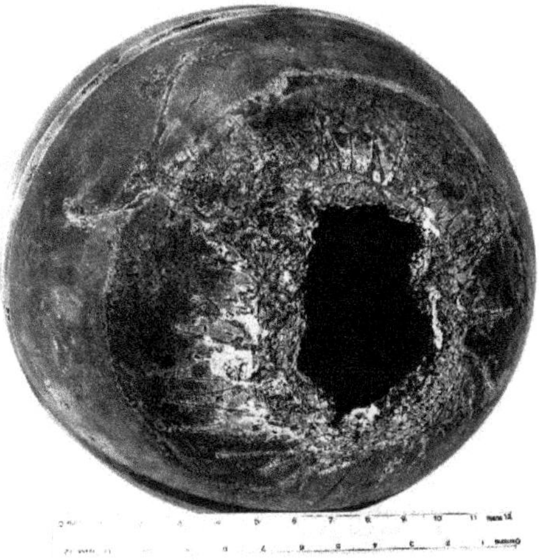

An Ashburton ball from the debris of Cosmos 482.

EXPRESS

In March 1995, Geoff received a quizzical phone call. The speaker said, 'How often do you read the *Ghanaian Times*?'[240] The caller didn't identify himself, but Geoff recognised the voice of Kettering Group member Stuart Eves. Geoff had previously asked Eves if he knew the fate of the German/Japanese satellite called Express. The Express mission was a repurposed Russian spacecraft with aerospace materials experiments on board that was scheduled to land in Australia. A two-week mission was planned but contact with the satellite was lost after two orbits; the Japanese rocket that Express launched on had malfunctioned and left the spacecraft in too low a trajectory. In his defence analyst role, Eves had come across an article in a collection of foreign space-related news reports that appeared to be a description of Express having crashed in Ghana. Eves sent the article to Geoff, but the poor quality of the fax made it almost unreadable. Based on data from *Aviation Week* and other sources, Geoff plotted the satellite's path and found it had passed over Ghana during its third orbit, around the time it had apparently vanished.[241]

At the Ghana High Commission in London a few days later, Geoff found the article in the *Ghanaian Times*. The satellite had Cyrillic writing on it but the Russian Embassy in Accra had denied any knowledge of the spacecraft. Geoff attempted to contact the newspaper editors responsible for the story but received no response. While pondering what to do next, he put out feelers for information about the missing craft. Several months later, he received a response that indicated someone else was about to publish the information that he had developed. He quickly wrote a summary and sent it to Jos Heyman, editor of the *News Bulletin for the Astronautical Society of Western Australia.*

Geoff chose the *News Bulletin,* he said later, to avoid offending US Space Command, which had refused to catalogue Express as an orbiting object, and because he was a member of the Astronautical Society of Western Australia and had previously published several articles in their newsletter. He may also have been thinking that, since Express was supposed to have landed in Australia's extensive Woomera rocket range, interested parties in that country might be more likely to read his analysis.

The article circulated in the aerospace community and was read by a member of the organization that was supposed to have recovered Express in the Australian desert. They contacted the German space agency DARA (now DLR) and said they knew where the missing Express satellite was. DARA sent a transport aircraft to Ghana to fetch the wayward spacecraft. Geoff subsequently received heartfelt thanks from the German space agency: 'I have enclosed [with] this letter some pictures of Express before and after the flight for you to remember this whole astonishing story in which you played quite a major role, since without you, the capsule would probably have stayed in Ghana forever.'[242] Stuart Eves noted, 'While most of the planned experiments failed due to the vehicle not reaching the intended orbit, the spacecraft also had external samples of potential re-entry materials, to determine how the materials would fare during entry; these experiments did yield results.'[243] The Express spacecraft was subsequently displayed at a Berlin airshow in 1996.

Express satellite before launch.

Express satellite after recovery.

11

Satellite Networks

1972 – 1997

NAVIGATION SATELLITES

In late 1972, Geoff published an in-depth analysis of a family of Cosmos satellites that shared an orbital inclination of 74 degrees. In the previous three years he had identified 37 launches, some carrying several satellites, and separated them into several subsets. He analysed each subset and made predictions—or admitted when he couldn't—about the possible purpose of each group. To do this, he collected missions into groups with similar orbital periods, then looked for other patterns, such as spacing of the orbital plane and apogee and perigee data.[244] He assessed one subset with a 105-minute orbital period to be the Soviet Union's first navigational satellite constellation, which he had identified with the help of his new programmable calculator. (See the table on the next page.)

The Kettering satellite trackers continued their detailed study of various Cosmos satellites looking for orbital characteristic similarities to help deduce their purposes. In 1976, Geoff and his 'distant cousin' who had previously worked in Fiji, Chris Wood ('my father's cousin's daughter's son'), analysed several Cosmos missions and, with the help of other clues, discovered the existence of a Soviet navigation satellite (navsat) system.[245] They had studied telemetry from the spacecraft and found timing signals that were consistent with navigation satellite functions; the orbital parameters and transmission frequencies were nearly

74-degree Cosmos satellites

Electronic Ferreting (95 min)	Unknown (100 min)	Doppler Navigation (105 min)	Unknown (109 min)	Military Communications (115 min)
269	292	378	256	336-343
315	304	381	272	411-418
330	332	385	312	444-451
358	371	422	409	Oreol*
387	372	465	426	
395	407	475	457	
425	468	489		
436	494			
437				
460				
479				
500				

Notes:

With a few exceptions, all spacecraft here are considered to be military payloads.

*Science mission

identical to the US' Transit navigation satellite constellation.

In 1978, the Soviet Union launched Cosmos 1000, a navigation satellite. Kenneth Gatland, a well-respected writer on space matters, described the first 1,000 Cosmos satellites in *New Scientist*, giving credit for detective work on the Soviet programme to Desmond King-Hele and Geoff Perry. Gatland noted that amateurs like Geoff 'have been able to place on public record many hidden features of Cosmos activity.'[246]

Some years later, Geoff told Patrick Moore that he thought the Kettering Group's most important achievement was the deciphering of the Soviet navigation satellite telemetry, presumably because it was the most difficult task that the Group attempted.[247] Using an upgraded strip chart recorder—a second-hand AEI Polygraph instrument—to display the satellites' signals, Geoff and Chris manually measured and analysed the 3,000 bits of data per minute received from the navsats. They observed that the satellites were sending time signals every second to receivers in the Soviet Union and updating their position in space every three minutes

Perry and Wood published a follow-on article in 1980 continuing their analysis of the Cosmos navigation system. Publishing in the prestigious journal *Philosophical Transactions of the Royal Society*, they updated previous articles identifying Cosmos spacecraft that transmit signals for surface navigation use.[248] Geoff and Chris described the complexity in the telemetry signals including how the satellite calculated its position and error correction methods. Chris Wood was the third overseas member of the Kettering Group, after Sven Grahn and Richard Flagg.

A researcher at the University of Leeds, Peter Daly, partnered with Geoff in 1986 to continue his earlier analysis of the Soviets' navigation satellite system. Writing in the journal *Space Communication and Broadcasting*, the authors described the navigation satellite constellation and how the spacecrafts' telemetry allowed surface receivers to determine their position.[249] They also noted that several bits of telemetry were undeciphered but might relate to satellite or data stream health. The two authors updated their findings a year later in the same journal covering developments from mid-1985 to mid-1987.[250]

In another piece, this time for *Space Policy* in 1997, Geoff updated previous Soviet navigation satellite articles and provided detailed analyses of navsat launches from 1986 to 1996. He concluded that the Russians were trying to optimise usage and on-orbit storage of satellites to reduce the launch load and conserve booster capability.[251] Geoff also provided information to *GPS World* magazine in 1997, updating the Russians' GLONASS (Global Navigation Satellite System, the Russian equivalent to the US' GPS) satellite launch information and navsat on-orbit operations. He analysed the current state of the GLONASS system for the same magazine the following year, discussing the system's future plans and potential use.

Geoff was also quoted in yet another front-page *New York Times* article in the summer of 1997 on the status of the Russian space programme. The article's author concluded that the Russian programme was underfunded and struggling to fulfil its promises, especially regarding the upcoming construction of the International Space Station. Regarding Russians launching foreign satellites to help with cash flow, Geoff said, 'Commercial launches [are] a big priority. They have not [cancelled] a commercial launch.'[252]

HUNTER-KILLER SATELLITES

Starting in the mid-1960s, Geoff was regularly contacted by reporters from around the world asking for comments on the latest Soviet space activity. In late 1968, the Soviet Union had launched Cosmos 248 into an unusual orbit, followed by Cosmos 249 and Cosmos 252. Evidence revealed later showed that Cosmos 249 and Cosmos 252 had both disintegrated in orbit while in the vicinity of Cosmos 248.[253] US experts suggested that Cosmos 248 may have caused the explosions of the other spacecraft. Analysts were concerned that the Soviet Union was developing anti-satellite capability to counter US space reconnaissance platforms.

In early 1970, the *New York Times* ran a front-page article about the Soviet satellite 'destroyer' capability, with quotes from Geoff concurring in general with the possibility. He concluded, 'It is not unreasonable to assume that the Russians have added a destruction capability in addition to interceptions.'[254] He was the only space expert quoted in the article who was willing to be named. In a declassified CIA memo, the Agency's Foreign Missile and Space Analysis Center commented on the *New York Times* story. FMSAC disputed the article's premise that 'experts believe' the Soviets have an anti-satellite capability. Referring to Geoff's more circumspect comments in the story, FMSAC noted, 'Mr. Perry seems to have his facts right, contrary to the *New York Times* article.'[255] Geoff had previously developed his analysis of Cosmos 248's activities in the second part of the 'Cosmos Programme' series in *Flight International.* He subsequently provided more details on his thoughts about Cosmos 248 in a letter to *Spaceflight* magazine in mid-1970.

Many years later, Mark Severance was interviewed by the CIA for an analyst position, in part because of his affiliation with the Kettering Group. He recalled one of his interviewers saying, 'The Kettering Group is really good. They don't get everything exactly right sometimes, but based on information that they have, their conclusions are either entirely correct or very reasonable for the dataset that they are looking at.'[256]

Continuing to analyse Soviet satellite characteristics to categorise the spacecraft by purpose, Geoff and his pupils noticed unusual properties of Cosmos 699. It was launched into a unique orbit, then made very small changes over time to its trajectory. Geoff and Russell Atwood, a pupil assigned to keep track of Cosmos 699, surmised that the satellite was a test bed for a new low-thrust rocket engine due to its unique orbital changes.[257] The satellite appeared to partially disintegrate after several months in orbit, resulting in dozens of fragments in Cosmos 699's orbital path. The Kettering trackers surmised that Cosmos 699 was part of a satellite interceptor system, and it had been destroyed upon completing its mission so it would not fall into an adversary's possession.

Charles Sheldon summarised the Kettering Group's Cosmos 699 findings in his 1976 Congressional report on Soviet space activities.[258] 'Geoffrey Perry and Russell Atwood at Kettering Grammar School studied changes in the mean orbital period of [C]osmos 699 and concluded that the spacecraft's mission involved trials of a micro-thruster which, over a period of months, moved the payload both up and down as if electric rocket tests were being carried out.'[259]

COMMUNICATIONS SATELLITES

The Soviets had sent Molniya satellites into a highly elliptical orbit supporting communications throughout the USSR. But until the early 1970s, they hadn't developed rockets powerful enough to get a communications platform into geostationary orbit. Their primary launch site, Baikonur, was so far north that inserting a satellite into the correct position above the equator required a large amount of energy to change the spacecraft's orbital plane, along with the thrust needed to lift a satellite to the approximate 35,780-kilometre-altitude (about 22,300 miles) orbit so the satellite remains stationary over a point on the Earth.

The Molniya Orbit

The Molniya orbits are unique and, when first introduced, were considered a novel solution for providing communication relays across the high-latitude expanse of the Soviet Union. The Molniya orbits are highly eccentric with apogees around 40,000 km (22,300 mi) and perigees near 600 km (375 mi). These altitudes result in an orbital period of approximately 12 hours.

The Molniya orbits are inclined at inclinations of 63.4 degrees. The choice of this inclination has two important effects. First, given the high apogee once the satellite reaches its most northern latitude (equivalent to the inclination of 63.4 degrees) it appears to 'hover' in place, providing the ability to relay communication across the Soviet Union for many hours. Second, at the inclination of 63.4 degrees, the effects of the perturbations on the orbit due to the Earth's equatorial bulge are nullified. This allows the satellite to repeat its ground track on a daily with only occasional minimal corrections from the firing of small onboard thrusters.

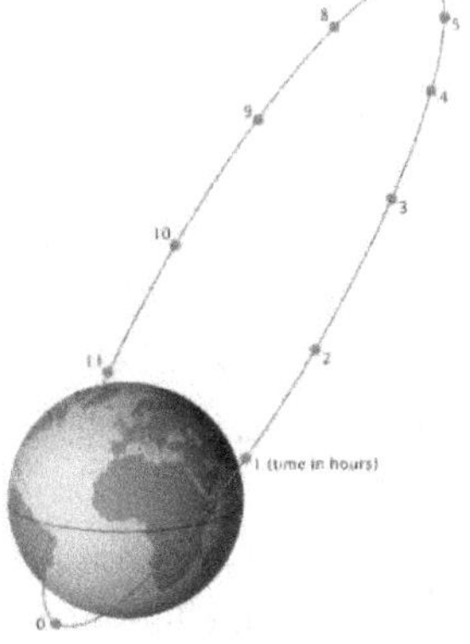

Typical Molniya communications satellite 12-hour orbit.

Geoff studied the orbital elements for Cosmos 637, launched in March 1974, and determined that it was a test of geostationary capability.[260] The Soviets had announced several years earlier that a constellation of geosynchronous communication satellites would be operational by 1970, but that had not occurred. Geoff concluded that Cosmos 637 was launched in support of this mission, but he had no estimate as to when the geostationary network would be put into service.

Geoff and one of his pupils, Philip Perkins, studied the Soviets' Molniya communication satellite constellation in the early 1970s, trying to ferret out its operational details. Geoff had just started receiving satellite data from NASA to supplement the RSRS and RAE information he subscribed to. The new NASA information allowed Geoff and his pupil to see a pattern in the orbits of the Molniya spacecraft.[261] The satellites were spaced around the world in pairs, 90 degrees apart in longitude to enable constant coverage over the whole of the USSR. They deduced that older Molniya platforms were regularly replaced by newer ones inserted into similar orbits, and that the satellites' orbital planes drifted at a rate to keep them tracing the same ground track throughout the year.

Geoff described the significance of the new NASA bulletins:

> In 1974 I began receiving regularly by airmail two-line orbital element sets for all objects currently in orbit. These are produced in NORAD's Cheyenne Mountain complex by US Space Command and are distributed to *bona fide* researchers by NASA's Goddard Space Flight Center near Washington, D.C. The [two-line elements, or TLEs] enable one to compute a satellite's position, velocity and acceleration at a particular instant and to predict its subsequent motion. Although we were receiving signals from only a few satellites, I realised that these [TLEs] were a valuable source of data from which deductions could be made about the purpose of various classes of Soviet satellites which we could not monitor. Individual pupils were allocated different sub-sets of satellites and by collecting the [TLEs], analysing plane-spacings, orbital periods, etc., they were able to infer the operational status of communications, navigation, missile early

warning and electronic intelligence gathering constellations.[262]

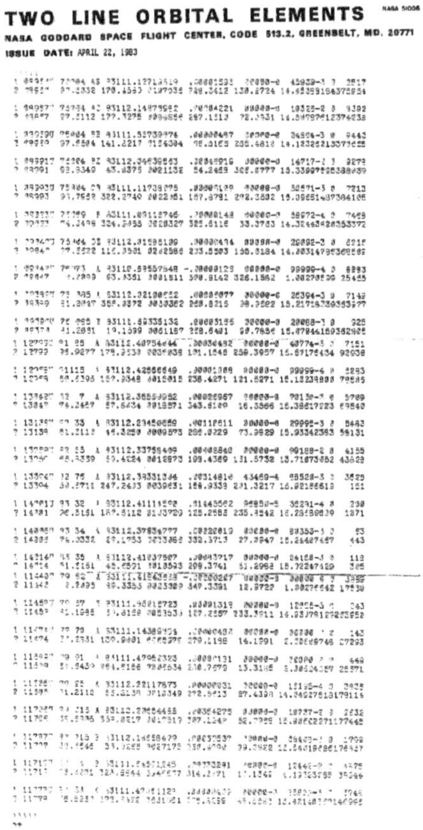
TWO LINE ORBITAL ELEMENTS NASA 51006
NASA GODDARD SPACE FLIGHT CENTER, CODE 513.2, GREENBELT, MD. 20771
ISSUE DATE: APRIL 22, 1983

A page of two-line elements (TLEs) distributed from the NASA Goddard Space Flight Center. The use of TLEs was the main method that the Group employed to track satellites and analyse orbital behaviour, especially after they started receiving the NASA bulletins in 1974.

Geoff followed up on his work on the Molniya communications satellites a few months later, noting that a new series of Molniya satellites had started replacing the older models. He also speculated on the purpose of several unidentified earlier launches, observing that, while they flew in highly eccentric Molniya-style orbits, they were inserted into orbital planes that didn't match up with the Molniya system. Geoff wondered, 'Are they military [communication satellites] or are they, as has been recently suggested by Dr. Charles S. Sheldon, carrying out early warning work on missile launches and nuclear tests?'[263] He was referring to commentary by Sheldon in a recent edition of *Air Force* magazine.[264]

Later, these uncategorised spacecraft were indeed identified as belonging to a new Soviet early warning satellite system called Oko (Russian slang for 'watchful eye') which flew in a Molniya-style orbit with apogees synchronised to latitudes corresponding with US ballistic missile launch sites.[265] The high apogee allowed for a long observation time at the desired latitude, because satellites travel slower at higher altitudes.

RADAR OCEAN RECONNAISSANCE SATELLITES

In 1978 Geoff and his pupils made one of their most consequential discoveries, although they didn't announce it. The Soviets were famous for saying nothing about a spacecraft until it was in orbit and functioning properly. Any mission that failed to reach orbit or did not achieve its goals was either publicly ignored or briefly acknowledged with a generic statement implying 'mission accomplished.' However, one mission's failure could not be disregarded or explained away. The infamous orbital decay and uncontrolled re-entry over Canada of the Soviets' nuclear-powered radar ocean reconnaissance satellite (RORSAT), Cosmos 954, galvanised the world's attention in early 1978.

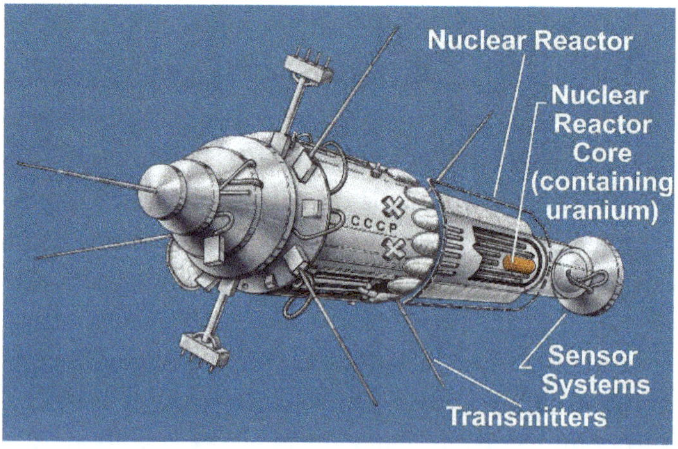

The Soviets' radar ocean reconnaissance satellite (RORSAT) configuration.

Two Kettering Group pupils, Stuart Ganney and John Kellett, had been analysing pairs of Cosmos satellites, watching them manoeuvre in tandem and then rise to a higher permanent orbit. Early on, the mission and architecture of the satellites was not known to Kettering observers. When Cosmos 952 and Cosmos 954 initially behaved the same way, they expected both satellites would be sent to high orbits and their missions terminated.

Ganney and Kellett handed over their monitoring activity to fellow pupil Andrew Driver, who expected Cosmos 954 to follow Cosmos 952 to a higher orbit by mid-November 1977. He noticed seemingly erratic orbital corrections for Cosmos 954 in early November, followed by a steady decrease in orbital period and hence altitude. He assumed the orbit would be raised at some point, and predicted that, if it wasn't, Cosmos 954 would enter the atmosphere in early spring 1978. Then, on 6 January 1978, the satellite's observed period began decreasing more rapidly, and Driver determined that the spacecraft could deorbit as soon as 25 January. 'Consequently,' Geoff wrote later, 'it came as no surprise when, in the afternoon of 24 January, [United Press International] phoned the school to ask if we "knew anything about this satellite that [had] fallen over Canada."'[266] Driver, at age 14, had predicted the deorbit of Cosmos 954 to within a day of its actual decay.

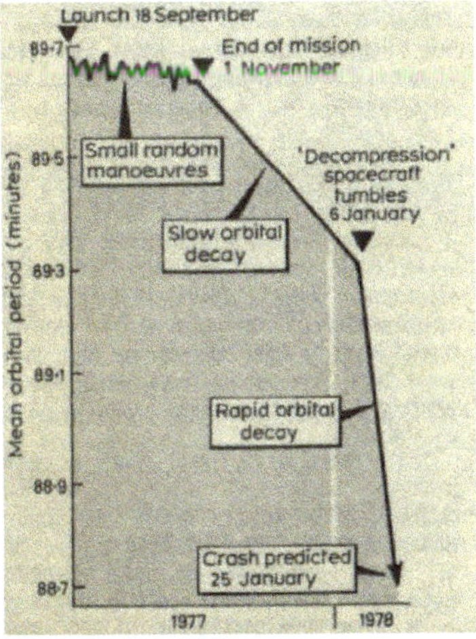

Kettering Group prediction of Cosmos 954's orbital decay.

RORSAT

Cosmos 954 was one of a series of radar ocean reconnaissance satellites. The Soviet Union needed a way to track US Navy warships throughout the oceans of the world. Conventional reconnaissance satellites did not have sufficient search abilities, nor did they have a wide enough field of view to cover countless thousands of square miles of water. The best solution was to fly a large radar that could sweep the ocean below it for ships big enough to reflect its radar beam. Such a satellite would need significant power for the radar. Using solar panels large enough to generate sufficient power would cause too much drag at the low altitudes necessary for ship detection.

A nuclear reactor was designed which could be separated into sections allowing for disposal of nuclear material when the mission was concluded. To avoid dropping nuclear fuel into the atmosphere, the expended radioactive material was to be boosted into a high orbit that would ensure the radiation would decay to safe levels before the satellite re-entered the atmosphere. Usually this altitude boost worked well, but in a few cases, like Cosmos 954 and two other satellites discussed below, the manoeuvre failed, and the satellite became a potential hazard to anything on the surface along its ground track.

As with any spacecraft's natural orbital decay—being susceptible to variable upper atmosphere conditions—predicting the exact time and location of Cosmos 954's re-entry was extremely difficult for the Soviets, the CIA and the Kettering Group.

The Soviet Union did not release any information on Cosmos 954 prior to or during the mission, other than the initial orbital parameters. The Kettering Group members had eventually determined the function of Cosmos 954 but had not realised the satellite was powered by a nuclear fission reactor. The Group was aware of Cosmos 954's fate well before the situation was made public. Geoff admitted some years later that 'this was the "scoop" that got away, as in the previous November, we had told no one of our conclusions, other than the close community of amateur observers of artificial satellites. The possibility of any part surviving re-entry and falling in a densely populated area had not occurred to us.'[267]

The CIA was also aware of the situation with Cosmos 954; they knew what the satellite's purpose was and how it was powered. Military space sensors had detected the spacecraft's odd behaviour in November; the intelligence community had been alerted and had notified the White House. A US Defense Department space policy advisor some years later wrote a classified description of CIA and White House activities related to the threat of Cosmos 954. His article, since made public, is as entertaining as it is informative, mixing lively writing with a short primer on the principles of intelligence gathering and exploitation. Notably, the CIA and the Kettering Group detected Cosmos 954's decaying orbital conditions at about the same time—during both the initial indications of trouble in November, and the apparent loss of satellite control in early January.[268]

Geoff and his pupils in late 1982 were analysing Cosmos 1426 because of its unique orbit. They were trying to determine the satellite's purpose, studying each set of orbital data as it became available. (Cosmos 1426 was found to be the first test of an electro-optical reconnaissance system for the purpose of an early missile launch detection capability.) Geoff couldn't get any data for the satellite over the New Year's holiday, but on 4 January 1983, while enquiring about Cosmos 1426, he received orbital data for a number of parts of Cosmos 1402, a satellite suspected of being an ocean reconnaissance platform, a RORSAT. Cosmos 1402 was assumed to be powered by a nuclear reactor and to be similar in function to Cosmos 954. As Geoff later recalled, that evening '[*Aviation Week and Space Technology* magazine correspondent] Craig Covault phoned at 9:26 pm. He had nothing new about Cosmos 1426, but in asking around, he had been told that the RORSAT was ready to come in like Cosmos 954.' [269]

Geoff and Sven Grahn traded information on 5 January indicating that Cosmos 1402 had split into three fragments, one of which was still at a low altitude and appeared to be the reactor portion of the satellite. Geoff called his contact at Independent Television News,

Frank Miles, that morning about the impending uncontrolled entry but asked Frank to hold the story until he confirmed it. That same day, the US Department of Defense announced that Cosmos 1402's nuclear reactor was expected to make an uncontrolled entry into the atmosphere by the end of the month. The satellite's reactor core had about 45 kilograms (100 pounds) of nuclear material that could pose a radiation hazard if it landed in a populated area.

After the US announcement, Geoff told ITN he estimated the spent nuclear fuel would re-enter the atmosphere on 7 February.[270] Group member Max White at the Royal Greenwich Observatory photographed Cosmos 1402 with a specialised camera for satellite tracking and orbit determination and provided tracking data to Geoff. Geoff's calculation was more pertinent than the US DoD's; the DoD estimate was for a relatively harmless piece of the satellite, while Geoff was concerned about a much denser piece that was the spacecraft's nuclear fuel container. Geoff spoke later that day to the *New York Times* about the satellite fragments losing altitude, saying 'it could prove very dangerous, but I don't want to speculate too much on that.'[271] The *New York Times* article about Cosmos 1402 quoting Geoff appeared on page one of the newspaper the following morning. Some 300 journalists contacted Geoff during the Cosmos 1402 fuel container's gradual descent from orbit over the following 32 days; Geoff's calculation of a 7 February atmospheric entry was accurate.

Early in the morning of 6 January, Ted Koppel, host of the US' ABC television news show *Nightline*, interviewed Geoff about Cosmos 1402. Geoff spoke live to Koppel from the ITN television studio at 4:30 am London time. Also participating in the interview was Marcia Smith, Congressional Research Service Senior Level Specialist in Aerospace and Telecommunications Policy, former colleague of Charles Sheldon, and a Kettering Group member. Koppel and the two analysts discussed the probability and possible impacts of radioactive debris falling to Earth. Commenting on how some parts of the satellite may survive re-entry, Geoff said, 'The largest piece that came down in Canada [during the Cosmos 954 crash] that I know about was over a metre long, and something like, I'm going to mix my units now, six or seven inches in diameter. They called it a stovepipe.' Koppel said, 'What do you conclude from that?' Geoff replied, 'I wouldn't want to be hit by a piece that size.' Koppel, laughing, said, 'I'm with you on that point.'[272] For a UK television audience later in January, Geoff discussed the impending re-entry with ITN correspondent David Chater.

On 8 January, Geoff commented in a *Daily Telegraph* newspaper article, 'The [Cosmos 954] satellite split into three component parts as normal, but instead of the nuclear reactor being raised to the safe orbit at 950 km [590 miles], it remained in the low orbit of 243 km [151 miles] to 231 km [144 miles] above the Earth. When [the Soviets] say this was always intended, I think it is a pack of lies.'[273] Later, Geoff wrote of that comment, 'To my horror I was reported accurately in the following day's edition with 'Pack of Lies' as a subtitle. I was greatly relieved when it was admitted that the satellite was out of control.'[274] The Cosmos 1402 spacecraft deorbited over the Indian Ocean on 23 January and its nuclear fuel store re-entered over the South Atlantic in early February, in line with Geoff's prediction. All the satellite's component parts were believed to have been destroyed by heat during descent through the atmosphere.

MILITARY OPERATIONS

Geoff was becoming known to a wider defence-oriented audience. He summarised the latest information on Soviet satellite systems for the US journal *Military Electronics/Countermeasures*. Though the article focused on Soviet ELINT spacecraft, Geoff included descriptions of most of the systems that interested him, including ocean surveillance, navigation, communication, early warning, meteorology and anti-satellite missions. He emphasised how some platforms had collected data related to the Falkland Islands conflict, which had just ended when the article appeared.[275]

In 1992, Geoff published an analysis of military uses of space during the 1990-91 Gulf War. He described the US, European and Soviet systems that were available at the time and how they were employed to monitor and shape the battlefield. Geoff concluded that the Gulf War 'can be claimed as the first conflict in which space systems played a decisive role.'[276] He acknowledged that the Iraqis had virtually no satellite support and that the Allied coalition significantly benefitted from such access. Geoff quoted a US official as saying, 'The space support for Desert Storm and Desert Shield will probably be the minimum support expected in any future crisis.'[277]

He wrote a short technical article in early 1994 for the *Journal of Photographic Sciences,* describing how various orbital parameters are calculated to explain their significance in placing satellites over specific targets.[278] He borrowed significantly from his 1972 paper on Cosmos coverage of the Indo-Pakistani war. Geoff also discussed ground track and related orbital period data for navigation, ocean observation and communication satellites.

12

Wide-Ranging Activities

1971 – 1991

Geoff was happy to share the Kettering Group's observations of Soviet satellites, both crewed and uncrewed, with any interested parties; tracking operations in the 1950s and 1960s being imprecise enough that independent sightings could significantly enhance orbital calculations. Along with sending results to the UK's Radio and Space Research Service, he regularly wrote to Academician Boris Petrov, chief of the Intercosmos programme, a cooperative space exploration effort between the Soviet Union and several Eastern bloc nations. In one exchange of letters, Geoff forwarded voice transcripts from several Soyuz missions and asked about listening to the first Intercosmos satellite.

In a letter translated by Kettering Grammar School Russian language teacher Cliff Brown, Academician Petrov, apparently misunderstanding Geoff's request, replied that the satellite's transmission frequency could not be changed without great effort. Geoff apologised for the misunderstanding, sent another collection of data, and described the nature of signals received by the Group from the second Intercosmos spacecraft. Later, a report of the Kettering Group's activities described how Geoff shared the Group's findings: 'When a satellite log is completed, the Kettering data is parcelled up and sent to the Russians so they can compare it with their own information, and, said Perry: "They always write back and say thank you."'[279]

The Intercosmos programme

The USSR sponsored a programme called Intercosmos, involving multiple countries in the Soviet space programme. Participants included Eastern Bloc countries, non-aligned countries like India and Afghanistan, and Western-aligned states including Japan and the UK. Both crewed and uncrewed missions were conducted from 1967 to 1994. Robotic missions in the programme studied the Earth's atmosphere, the Sun and space physics. Twenty Intercosmos crewmembers participated in research on board the Salyut 6, Salyut 7 and Mir space stations.

In addition to his own projects and the Kettering Group's activities, Geoff had become a follower of burgeoning space programmes around the globe, and he couldn't resist setting the record straight when he had an opportunity. Geoff read an article in the journal *Physics Bulletin* that he felt mischaracterised the programmes of many of the world's space-faring nations; he wrote a pointed reply. The article's author asserted that the US had launched the greatest number of satellites in the world during 1970. But Geoff knew that wasn't true and took the author to task, citing numbers showing the Soviet Union was the leader in launches.[280] He also refuted a notion that the first Chinese satellite was not much more than a stunt and lamented the article's omission of the first Japanese satellite.

Rumours about a Soviet human mission to the Moon had been circulating in the West for years. Though Soviet officials denied such efforts existed, Western analysts found ample evidence of an active lunar programme behind the Iron Curtain. Several cosmonauts had earlier indicated that they had expected to land on the Moon before the United States did. In the late 1960s, the CIA had gathered data on a possible Soviet Moon programme and made predictions—some good and some inaccurate—based on the incomplete information that they had covertly shared with NASA.

In a *New York Times* front page article in 1971, speculation ramped up concerning several Cosmos missions that appeared to be tests of lunar mission hardware. Four Cosmos satellites had been identified as possible testbeds of propulsion and manoeuvres equivalent to the capabilities of the Apollo spacecraft that the US had flown to the Moon. Geoff, described as the 'British space expert, who alone would speak for attribution,' was quoted in the article saying, 'The Soviet Union is testing a large rocket engine similar to the one aboard Apollo.'[281] The story also cited American space experts who concurred with Geoff's analysis, but those experts remained anonymous, perhaps because their information or sources were classified.

Through his accomplishments in satellite observing, Geoff had been invited to join the Optical Tracking Subcommittee, which reported to the British National Committee for Space Research. Desmond King-Hele was a member of this subcommittee, and later chairman. The subcommittee was initially comprised of about 50 skilled satellite visual observers who regularly sent their observations and data to the RAE and the RSRS; the observations were considered 'so valuable in determining satellite orbits for subsequent analysis in geophysical research.'[282] Prospective Kettering Group member Max White knew that Geoff would be at the 1980 meeting of the Optical Satellite Tracking Subcommittee and was keen to meet him. White recalled later, 'I made sure I sat next door to him on the coach when we came down [to the meeting].'[283] White was an active member of the Kettering Group for the next 20 years. Other Group members at the meeting included Andrew Driver, Dave Hawkins, Peter Wakelin and Pierre Neirinck.

Max White worked at Aston University, then at the Royal Greenwich Observatory running the Baker-Nunn satellite tracking camera, the state-of-the-art optical tracking system at the time. He supported production of satellite prediction bulletins using data from NASA and RAF Fylingdales as well as the RGO cameras. His specialty in the Kettering Group was keeping track of Soviet tracking ships that deployed around the world in support of Soyuz crewed missions. White wrote an appendix to Marcia Smith's first volume of the *Soviet Space Programs:*

1981-87 report (more on this series of reports in Chapter 15), describing how the tracking ships supported Soviet space activities.

White commented later, 'When Geoff did a lecture tour, he used to give me the fax numbers of where he was so I would fax him the latest information…so that when he gave a talk, he could say "the latest Russian launch was at 10 o'clock this morning." So he was absolutely up to date.'[284] White added, 'I would ring him up at any time of the day or night—he literally said, "ring me up any time." There was one time I rang him about three in the morning. Literally after two rings he answered, I said, "My God, you were up," he said, "Oh, we're just about to drive to Bude for holiday."'[285] Geoff gave a talk entitled 'Listening to Satellites' during a 1987 meeting of the Optical Satellite Tracking subcommittee, and 'regaled the audience with a medley of interesting radio signals from recent satellites caught during his untiring pursuit of transmissions.'[286]

The British National Committee for Space Research

The British National Committee for Space Research (BNCSR), a unit of the Royal Society, was formed in 1958 from Britain's IGY Groups as the main UK interface with the newly established international body, the Committee on Space Research (COSPAR). Geoff and several Kettering Group members were active participants in the subcommittee, regularly sending in their observation data to Farnborough, where the main committee was headquartered.

COSPAR was and still is the international body that coordinates space research efforts around the world. All artificial satellites are given a COSPAR ID number and catalogued for reference. The Optical Satellite Tracking Subcommittee, along with the entire BNCSR, was dissolved by the Royal Society in 1989 as a cost-cutting measure. The subcommittee, renamed the Optical Tracking Working Group, was reinstated in 1990 after an emphatic plea by its membership, and it functioned for several more years.[287]

Reflecting Geoff's extensive knowledge and interest in the world's space programmes, he defended US astronaut Joe Allen in a letter published in *New Scientist* responding to a review of Allen's recently published memoir. The reviewer had described Allen as a 'rookie astronaut'. Geoff took the reviewer to task, pointing out that Allen had been an astronaut for 18 years, supported the Moon landing crews, served on a White House committee, directed a division at NASA Headquarters, and had flown on the Space Shuttle's fifth flight.[288]

Geoff was not afraid to argue against what he considered faulty information or conclusions, no matter what the source was. Along with a satellite tracking colleague, he composed a letter for *Aviation Week and Space Technology* discussing Soviet satellite longevity. The US Secretary of the Air Force, Edward Aldridge, had derided Soviet satellites because 'they don't last as long [as US spacecraft], they're not as high technology and their launch rate is, by necessity, much higher.'[289] The two authors countered this claim, presenting tracking evidence that older Soviet platforms could be reactivated and presumably still be useful in their mission. They hoped their evidence would 'sound a warning to those who take comfort from views such as those expressed by Aldridge.'

Geoff's work was often cited by internationally recognised authorities. In 1989, Australian defence and security expert Dr. Desmond Ball wrote a study of Soviet signals intelligence (SIGINT) capabilities. He had previously consulted with the US government on nuclear war plans, and his advice had been praised by former President Jimmy Carter. In his study, Ball cited Geoff's papers on Soviet ELINT networks and their ocean surveillance systems.[290]

On a late summer evening in 1991, Geoff stepped outside his home in Bude, Cornwall, to observe the Russian space station Mir and the Space Shuttle *Discovery* orbiter flying the STS-48 mission; the timing of two spacecraft passing overhead was coincidental as the vehicles were in different orbits. While making notes on the spacecraft moving across the sky, Geoff and his colleague Howard Miles, who was observing separately, noticed a thin white streak in front of *Discovery*.[291] Geoff left a message enquiring about the phenomenon for Mark Severance, who was working in Space Shuttle mission operations in Houston, Texas. Severance replied that, at the time of observation, *Discovery* had been dumping excess wastewater. The water had crystallised and reflected the Sun's light, resulting in the unusual sight; the streak had appeared in front of the vehicle because the dump had occurred through a vent port on *Discovery* which was pointed in the direction of orbital travel.

Part III

Impact and Legacy

13

Television, Radio and Print

1965 – 1993

Geoff and his pupils' activities received near constant media attention from the 1960s onwards. While Geoff published well over 100 articles and journal pieces, he was quoted and consulted for many more stories and videos the world over. Because his methods and findings were unique, Geoff was often the subject of profiles in major magazines and newspapers, as well as in television and radio programmes.

Kenneth Owen, the *Flight International* reporter who visited the school in 1964, published a profile of Geoff and a description of the Kettering Grammar School Satellite Tracking Group in a July 1965 edition of the magazine. Details of their activities were revealed to an international audience for the first time. Kettering High School pupil Barbara Jacques was featured, along with several boys from the Grammar School. Geoff's succinct comment in the article was, 'We've got a lot more out of [satellite tracking] than we ever imagined.'[292] Owen noted in the story, '[Barbara Jacques] became a keen member of the school's space group, has just taken her A-level exams in physics and mathematics [similar to Advanced Placement exams in the US], and intends to become a Junior School maths teacher.'[293] Later, Geoff would observe, 'Concentration on a sub-set of Cosmos satellites which disappeared from orbit after flights lasting only eight days or less—long before they would have decayed naturally—was the key to unlocking many of the secrets of the Soviet military space programme first carried out under the Cosmos label.'[294]

Throughout 1966, Geoff and his pupils were attracting attention from the media. In early March, British newsman Lionel Hampden came to the school and interviewed Geoff and several pupils for Associated Television (ATV), which at the time was part of the Independent Television Network. They demonstrated their tracking abilities to the reporter by monitoring the Cosmos 110 satellite, which was carrying two dogs. Hampden elicited this comment from Geoff: 'When we started off, we were mainly after one recording for one experiment, but it's grown and you can't really tell what is going to come out of it. We've measured the density of the air at 135 miles altitude. We've heard satellites as they've been recovered. We just can't tell. That's the beauty of it.'[295]

Michael Sinnett, who was interviewed for the report, remembered being excited about appearing on the news programme. 'I rushed home to ensure that my family were ready to watch it. It was very strange, yet exciting, to see myself on television for the first time. I wondered how many of my friends had seen it or read the headlines in the newspaper. When I walked into Kettering town centre the following day, I felt like a film star and I expected lots of people to recognise me. Needless to say, this didn't happen!'[296]

Shortly before the flight of *Apollo 8* to the Moon in December 1968, the satellite trackers from Kettering gathered in the school's lecture hall. Across the Atlantic, acting NASA Administrator Thomas Paine was linked live to the Group members by phone. Several Kettering pupils asked questions of NASA's leader in Washington, D.C.[297] The interview aired on BBC a few days before *Apollo 8* astronauts Frank Borman, James Lovell and William Anders, who all appeared in the telecast in pre-recorded segments along with US rocket scientist Wernher von Braun, made their historic journey to the Moon. Geoff's colleague John Marshall took photographs during the programme's broadcast; the photos were displayed on the wall of Robert Christy's parents' house for many decades. Kettering Group pupils shown on the broadcast included Graham Coleman, Robert Christy and Arthur York.

The UK populace appreciated the ingenuity and effort of the Kettering Grammar School's satellite trackers. In the *Daily Mirror* newspaper, columnist Christopher Ward imagined Kettering pupils phoning a newspaper to say that they had 'just launched Britain's first unmanned probe to Mars and Mr. Perry, the physics master, wants you to be the first to know.'[298] The caller explains that they had 'stayed behind after school for six months building the rocket in the school metal shop and preparing the liquid oxygen in the chemistry lab.' The pupil must hang up because 'Mission Control wants me back in the cricket pavilion right away'—he must go to exercise class.

One of the science reporters that Geoff regularly worked with in the 1960s, Peter Fairley of the *Evening Standard*, moved to the Independent Television (ITV) network as their science correspondent. In the summer of 1971, Fairley invited Geoff to participate as an expert commentator during ITN's coverage of the Apollo 15 mission to the Moon; he and Fairley would work together for many more years.

In the summer of 1975, Geoff participated in Independent Television News broadcasts in the UK which covered the Apollo-Soyuz Test Program crewed missions. ASTP was a joint spaceflight conceived during the US-USSR 'détente' relations of the early 1970s. It was the first internationally crewed human spaceflight and, in the view of many, the end of the US-Soviet 'space race.' The mission successfully docked an American Apollo spacecraft with a Soviet Soyuz spacecraft, testing a common docking system and fostering cooperation as symbolised with a handshake. Geoff appeared with ITN reporters Peter Fairley (science correspondent) and Reginald Bosanquet (chief newsreader) in the studio, while Derek Slater and the satellite-tracking pupils supported Geoff from Kettering Grammar School.

Geoff appeared live on several ITN reports which covered key milestones of the mission, including the docking, the handshake and the landing of the Soyuz spacecraft. Real-time data from the spacecraft was fed to him live from the Kettering school laboratory via two dedicated telephone lines—one for voice transmissions and one for the shortwave telemetry. He described the signals that were being received and explained how anyone could listen to telemetry and voice signals as the spacecraft passed over the UK. Geoff also demonstrated a unique piece of hardware that allowed real-time tracking of the spacecraft's position in orbit. As he explained during a broadcast, '[The Soviets] have a globe inside their spacecraft that shows the cosmonauts exactly where they are over the Earth at any instant. I said to a retired engineer friend of mine, I said, 'Frank [Lewin], I could do with a globe just like that so we could see where the spacecraft we're tracking actually is.' About a year later he came up with this.'[299]

Orbital tracking device built by Geoff's friend, Frank Lewin.

From orbital data, Geoff had calculated the position of the Soyuz spacecraft as it revolved around the Earth throughout its mission. Whilst in the studio, Geoff refined his data using his daughter's Hewlett Packard HP45 pocket calculator. Frank Miles, ITN Associate Producer and Science Editor, used Geoff's most up-to-date data to create a graphic which showed the detailed ground track at the expected time of the handshake; the ground track indicated that the spacecraft would fly over the English Channel just south of the town of Bognor Regis. In the lead-up to this momentous event, the news generated much excitement in the town. One local restaurant invented a new drink, the Space Shake, which was described as powerful enough for the imbiber to 'be put into orbit.'[300] According to Charles Carmichael, a news photo circulated showing Bognor Regis town officials looking up into the sky as if they might see the handshake from the ground. In the event, the meeting between the crews was delayed and the actual handshake happened while flying over northeast France.

As part of the coverage, an ITN film unit visited Kettering and conducted interviews with the team there as signals were coming in. Geoff delayed his departure from London when ITN asked him to stay on for a couple of days after the Soviets announced the Soyuz spacecraft landing would be broadcast live.

A *Financial Times* television reviewer criticised the BBC's coverage of the ASTP mission, while complimenting the ITN ASTP reports. He praised Geoff's commentary as being 'blunt and good natured' and easily understood.[301]

ITN's Peter Fairley added his assessment of Geoff's studio work in a letter to him afterwards. 'Without a doubt the strongest memory of this [mission] will be your being around—with those "simple" but nevertheless quite fantastic calculations which helped put ITN in front again. The *Financial Times* were so right in singling out your performances for praise. For me it was great having you around—and if you felt any regret on leaving us, believe me it was echoed within the space team generally. We missed you. Thanks, Geoff, for making it all such great fun.'[302]

New Scientist magazine, in May 1978, featured a three-page article about the 20 years of Geoff's satellite tracking activities and his work with his Kettering pupils. Along with describing the highlights of the Group's work, Geoff had a lot to say. He noted, 'I feel if you are an active science teacher, you've got to have some line of research going along to keep your mind tuned.'[303] Recalling the tracking of Voskhod 2 in 1965, when cosmonaut Alexei Leonov performed the first spacewalk, Geoff said, 'By this time, we were getting pretty much involved in following anything Russian.'[304] Lamenting that the Group had to tape over old telemetry and voice communications for lack of recording media, he said, 'We were poor then. We are still poor.'[305] Describing how the pupils calculated predicted launch times, Geoff explained, 'It's very easy. We do the same as the Russians do. There's only one minute of the day you can go to match the (Salyut) orbital plane.'[306]

In the spring of 1980, Geoff starred in another profile article, this one in the *New York Times*. The author described him as the quintessential English 'gifted amateur.'[307] In the article, Geoff remembered calling Reuters from his holiday location in Bude, England, to report the launch of a crewed mission to Salyut 6 that the Soviets had not yet announced. Reuters sent out a worldwide story about the launch based solely on his report.

On his philosophy of instruction, Geoff said in the profile, 'I try to teach physics as a subject that has an impact on everything in our lives. Every boy here starts studying physics at age 11, and it clearly helps that I can relate simple principles to space projects which catch their attention. I was [teaching] mirrors this morning, and you can bring in parabolic antennas and radio telescopes very easily.'[308] Geoff mused about the rest of his career, 'I would like to work at a university training future physics teachers. I think I could do that. But there are no university jobs. I shall grow old here and retire quite happily to Cornwall in a few years.'[309]

Frank Miles contacted Geoff in March 1981, asking him if he would like to appear on the television news programme *60 Minutes*. Geoff agreed, and filming took place over several sessions. *60 Minutes* correspondent Morley Safer interviewed Geoff and some of the Group members. Separately, Congressional researcher Charles Sheldon was filmed at his office in Washington, D.C. The report was broadcast in February 1982, featuring Geoff, Charles Sheldon and pupil Andrew Driver. At the end of the report, Safer asked Sheldon, 'Do we ever rely on Perry for a bit of original intelligence that we don't already know?' Sheldon said, 'The answer to that is yes, and for reasons of government policy, I can't be specific about my examples. But the answer clearly is yes in my own experience.' Safer said, 'The reason you can't be specific quite simply is that it would show the Russians what we didn't know until Perry told us.' Sheldon replied, 'That's correct.'[310]

During the last year of his teaching career, Geoff had yet another profile article published about him, his pupils and their activities, this one in *Science Digest*. Along with a description of the highlights of the Kettering Group over the years, there were some illuminating comments. At that point, Geoff had been listening to satellites for 24 years, yet his enthusiasm seemed as boundless as ever. His sense of humour was evident, too. He listened to shortwave radios not only in school, but also at home and in his vacation caravan by the sea. 'One needn't be very far from a Russian satellite, need one?' he quipped.[311] Some years back, Geoff had heard cosmonauts transmitting their votes for the then-Soviet Premier Leonid Brezhnev. In the article, he remembered thinking at the time, 'I'd have voted for him too if I wanted to come back down.'[312] Kettering Group members Mark Severance and Richard Flagg were also featured in the *Science Digest* story.

In the 1980s, *Reader's Digest* claimed to be the 'world's most read magazine', printing over 30 million monthly copies in 40 countries around the world. The October 1983 UK edition featured a flattering profile of Geoff and the Kettering Group, perhaps the single most-read article ever about the satellite trackers and their inspiring leader. The story was reprinted in several other editions of the magazine, including versions in the Netherlands, Germany, India, China, Sweden and Australia. Also mentioned in the story were Andrew Driver, Robert Christy, Sven Grahn and Richard Flagg. Geoff ended the story with the pithy comment, 'The only way to close us down at Kettering would be for [the Soviets] to adopt a policy of complete openness on all matters appertaining to the uses of outer space—and we think that most unlikely.'[313] Unlikely then perhaps, but in a few years, increasing Soviet openness would start revealing information that Geoff and the Kettering Group had never imagined.

Geoff had been a space expert on ITN broadcasts since 1971. After he retired from teaching in 1984, he signed a contract with ITN to work exclusively for them as their space consultant. He regularly appeared on news programmes and wrote a bimonthly summary of space activity for the network. When Geoff retired from ITN in 1997, Peter Fairley, who recruited him to the network, wrote to Geoff, 'Making contact with you when I was at the *Evening Standard* was, quite truthfully, one of the most valuable events in my working life—both for the paper and for me. At ITN, your authority became our authority. We could be sure that what we reported or announced was right. You were not just a superb educator but a real communicator and there must be scores of young men who owe their interest in space to your enthusiasm.'[314]

We've already seen how Patrick Moore, presenter of BBC's *The Sky At Night* television programme, keenly followed Geoff's activities for many years, often mentioning the school's tracking projects on his television and radio broadcasts. One of his radio shows in 1985 was dedicated to Geoff and the Kettering Group members. Geoff spoke extensively during the programme, describing the Group's activities over the previous 20-plus years. He summed up the activities that he and Derek Slater had started: 'It was just two people's hobby, and the Group has gradually evolved over the years until it now runs me.'[315]

During the radio broadcast, Moore also spoke with Kettering members Mark Severance, Richard Flagg, Marcia Smith, Sven Grahn, and Robert Christy. Christy remembered how the Group operated: 'One fortunate thing was the Russian spacecraft that we were tracking tended to be launched in the middle of the day our time and they wanted them over certain parts of the Earth at certain times of the day. So we could get away with most of the tracking during normal school hours without disrupting things too much.'[316] Richard Flagg, the first US Group member, recalled his early days. 'Tracking satellites can be rather a lonely hobby if you're doing it all by yourself, and you certainly seek out other people that have the same interest… Having read about Geoff and the Kettering Group in the late 1960s I wrote to Geoff and told him about the satellite tracking I had been doing since Sputnik 1. We have been corresponding ever since.'[317]

Sven Grahn, speaking about Geoff during the programme, said, 'Quite honestly, every young man needs a mentor and Geoff has been that to me, and certainly encouraging me all the way. So what we've done together has been of immense value to my career as an engineer.'[318] NASA engineer Mark Severance said, 'I think Mr. Perry's whole motivation was to teach his students the methods of scientific research. It certainly taught me that. I think it's been a very directing force in my career; in a sense it's made my career.'[319] Researcher and Kettering Group member Marcia Smith said, 'There are other individuals who follow the Soviet space programme as a hobby perhaps, but there are none that would rank in the same category as the Kettering Group.'[320]

In a 1986 profile of 'amateur space sleuths' in the *New York Times*, the writer acknowledged Geoff's influence. 'The method and ethical rules of amateur trackers originate primarily with the founding father of the movement, Geoffrey Perry, a retired British physics teacher.'[321] Other amateur trackers from around the world described their tracking in the article, sounding very much like the Kettering Group members talking about their activities. In the profile, Mark Severance said of the Group's efforts, 'It's like working on a big jigsaw puzzle, only there are a lot of pieces, and they're buried.'[322] Geoff identified one of the reasons that the Kettering Group focused on classified activities. 'If we can find these things out, then they can't be secret, and that gives both sides a chance to talk openly about it.'[323]

The UK's Channel 4 in March 1987 broadcasted a programme called *Sputniks, Bleeps and Mr. Perry*. The one-hour docudrama included extensive remarks from Geoff, along with comments from Michael Sinnett, Derek Slater, Marcia Smith, and *Aviation Week* senior editor Herbert Coleman. Seminal events re-enacted include Geoff meeting Derek, the pair recording Sputnik IV's signal, showing pupils how to listen to satellites, and how Geoff figured out the location of Plesetsk. A British television reviewer described Geoff in the show as 'a comet of a teacher, pulling along pupils like particles.'[324] The reviewer noted Geoff's teaching style as 'sneaking in science by stealth.'[325] Geoff said in the programme that satellite tracking is 'an overriding passion—one's whole life, meals, holidays, sleep, were all dictated by satellite passes over Kettering.'[326]

After Geoff saw the docudrama, he said, 'It's not as bad as I feared, though it's not at all what I expected.'[327] Derek Slater was more complimentary: 'I think they've done very well. I certainly enjoyed it.' The show was originally part of Channel 4's *Equinox* science series but was pulled out to be a separate piece. Two years later, the US Public Broadcasting Service aired an edited version of the programme as part of its NOVA science programme series, titled 'The

Schoolboys Who Cracked the Soviet Secret.'

Michael Sinnett, who appeared in the show, later remembered, 'I thought they did very well [capturing people's personalities]. My mother was amazed because the lady who was playing my mum actually rang her up several times before to get to know her so she could play her properly. All the opening scenes were exactly as we talked about [with the television show interviewers].'[328] Michael's older brother David, a Kettering Group member in the early 1960s, had been filmed but his scenes were not included in the broadcast.

In 1989, Yorkshire Television asked Geoff to film a show that would be part of a series called *Science Fiction*—science stories that sounded a bit like fiction. Other entries in the series included a show about cold fusion, one detailing the first major computer hack, and an episode describing the discovery of pulsars, stars that send out bursts of radio waves. Geoff starred in an episode called 'The Crash of Cosmos 1900.' The previous year, the Soviets had apparently lost contact with their latest radar ocean reconnaissance satellite; there was concern over the potential re-entry of the spacecraft's nuclear reactor, as had occurred with Cosmos 954 in 1978 and Cosmos 1402 in 1983. This spacecraft, Cosmos 1900, had completed its mission, and as with the previous mishaps, it was waiting to be boosted to a higher orbit where its nuclear reactor power source would decay naturally and not be a hazard when it finally entered the atmosphere.

Geoff and some of his fellow trackers had expected the satellite to be raised to a higher orbit, and when that hadn't happened, they became concerned that another unguided entry with nuclear material might be imminent. Geoff contacted British authorities with his concerns; a few days later, Soviet officials admitted they could not control the satellite. However, unlike the previous satellites, Cosmos 1900 was equipped with a backup safety system which successfully ejected the nuclear reactor core into a higher orbit later that year. The television episode included some dramatization, but it was mostly Geoff explaining the story of Cosmos 1900. In a review of the episode that Geoff sent to the series' producers, he indicated he was pleased with the finished product (unlike his ambivalence toward the 1987 Channel 4 programme), and quoted Derek Slater's opinion that the show was 'brilliant and captured the real Geoff.'[329] The programme aired throughout England in March and April 1990.

Interest in Geoff's work continued worldwide into the 1990s. The Smithsonian Institution's National Air and Space Museum published a profile of Geoff and several Kettering Group members in the June/July 1990 edition of *Air & Space/Smithsonian* magazine. Geoff was shown happily retired on the North Cornish coast and still actively pursuing satellite tracking. The author described Geoff: 'He punctuates his talk with imitations of beeps, buzzes and Doppler shifts. He makes absolutely no small talk, nor does he indulge in false modesty.... He seems to view the Kettering Group's achievements, and his own, as triumphs of common sense, patience and the amateur ethic—a victory for schoolteachers and schoolchildren everywhere.'[330]

Several Group members were also featured in the article. Les Currington specialised in receiving Soviet weather satellite photos with his home-made reception rig. David Hawkins demonstrated how he tracked the latest photo-reconnaissance satellites. Long-time member Max White enjoyed monitoring VHF frequencies for Soviet cosmonauts' voice communications. When asked how he could still be enthusiastic after so many years of listening to telemetry and voices, Geoff said, referring to the mission that had launched that morning, 'But I've never heard Soyuz TM-8 before, have I?'[331]

In the late 1980s and early 1990s, the Soviet Union, and then the Russians, had been looking for ways to increase the productivity of farms and factories. Studies were conducted a study to see if it was feasible to use reflected sunlight to brighten early mornings and late evenings, thus encouraging people to work longer. The result was a test satellite called Znamya (Russian for 'banner'), a reflective disk that was to be deployed from a Progress cargo resupply vehicle after it had undocked from Mir space station. In 1993, Znamya 2's 20 metre (65 feet) mirror opened successfully and directed a five-kilometre (three mile) spot of light across parts of Europe and Asia.

Geoff reported on the Soviets' solar reflector for ITN, explaining that a practical mirror would have to be much bigger and positioned so a selected area of the Earth's surface was continually illuminated: 'You would have to have your reflector at geostationary distance, which is [approximately] 40,000 kilometres [approximately 22,300 miles] above the equator, and it would have to be very, very much bigger. You're talking there instead of a reflector 65 feet [20 metres] across, you're talking about something which is several miles [5 to 10 kilometres] across.'[332] A second test of Znamya with a bigger mirrored surface failed several years later, and the project was subsequently cancelled.

Znamya satellite in orbit near the Mir space station.

Robert Christy published a summary of the Kettering Group's activities in a 2014 edition of *Space Chronicle* magazine. He described in detail some of the Group's early activities and how classes of Cosmos missions were identified. Christy noted towards the end of the article, 'Geoffrey Perry and Derek Slater succeeded in their original aim to introduce study of satellites and space to enhance their students' learning experience, but the reality was demonstrably different from the original idea of studying orbits and the Doppler effect. Pupils learned to collect data, organise it, analyse it and publish the results and conclusions—exactly what scientists are expected to do.'[333]

14

Travels

1976 – 1993

Geoff's time and opinions were in constant demand. Not only did he personally answer every letter sent to him—often at length—but he regularly gave talks to local schools, universities and civic groups. He loved to travel and went abroad frequently, sometimes combining holidays with lectures and conferences.

In the summer of 1976, Geoff, accompanied by his daughter Isabel travelled to the US. They stayed with Charles and Jean Sheldon for two weeks in the Washington, D.C. area. As well as making a number of visits to the Smithsonian's National Air and Space Museum on the National Mall, Geoff met many colleagues with whom he had only corresponded before, along with luminaries like *Apollo 11* astronauts Michael Collins and Buzz Aldrin, Apollo veteran and future Shuttle astronaut Joseph Allen, and rocketry pioneer Frederick Durant III.

At NASA headquarters, Geoff and Isabel watched the touchdown and first photo transmissions from the Viking I Mars lander. Whilst in Washington, D.C., Geoff was a guest speaker at a National Space Club luncheon and lectured at the University of Maryland, with both addresses being well received. Geoff and Isabel also visited Kettering Group member Richard Flagg in Florida and toured the Kennedy Space Center, where Isabel snapped a photograph of her father dressed in an Apollo-era 'Snoopy cap' communications headset and a space suit glove which were on display, together with his Omega Speedmaster watch (as worn by Apollo astronauts)—a present from his wife Jean.[334]

Geoff at NASA Kennedy Space Center in Florida.

Due to Geoff's international reputation, he received an invitation from the Astronautical Society of Western Australia to deliver a talk in Perth. He participated in the August 1979 Australian Astronautics Convention and gave several presentations summarizing the activities of the Kettering Group. Geoff also gave a 'light-hearted' lecture to the ASWA entitled 'An Evening With Geoff Perry', where he told stories of 'eventful and unusual' occurrences from the Kettering Group's history.[335] Geoff was made an honorary member of the Astronautical Society of Western Australia. He was invited back to Australia in 1986.

His second trip was more extensive, ranging from Perth to Brisbane, Sydney, Canberra, Melbourne and Adelaide. He gave lectures and interviews everywhere he went and, by one account, 'at all his talks, his audience came away enthused over space happenings and fascinated at the very significant and positive deductions that can be made from continuous monitoring and recording of relatively simple data.'[336] His trip was chronicled by numerous Australian newspapers and television stations. One of the Australia trip organisers, Jos Heyman, visited London in 1995 and gave a lecture to the British Interplanetary Society with Geoff as the host.

During his 1979 Australia visit, Geoff presented a paper to the Australian Astronautics Convention titled 'The Soviet Space Programme: A Kettering View' which was later published in the Convention proceedings. The presentation covered major findings and activities of the Kettering Group, including their studies of Soviet reconnaissance and navigation satellites, meteorology platforms, communication satellite systems, and cosmonaut voice transmissions. The same paper, updated with the latest data, was presented to the Institution of Electrical Engineers meetings in central England in 1979 and 1980. Geoff then rewrote the paper to include more technical details and published it with the same title in the journal *Electronics and Power* in 1982.

In January 1980, the BIS held its first annual Soviet Technical Forum, later called the Soviet/Chinese Technical Forum. BIS publications had regularly covered Soviet space activities, and interest was high in the society for an in-person meeting of Western experts to discuss related topics. Kettering Group member Dave Hawkins gave a presentation about visual observations of Soviet spacecraft, and Geoff Perry attended the meeting. Nicholas Johnson, a US expert on Soviet space activities, also spoke at the 1980 forum about trends in the Soviet photo-reconnaissance programme. Johnson updated his presentation a year later in the *Journal of the British Interplanetary Society*, citing Geoff as a primary source of information for the article. Geoff gave his own talks at the Soviet/Chinese Technical Forum over the years and attended regularly. His last presentation was in June 1999, just a few months before he passed away. At the forum in 2000 the attendees paid tribute to Geoff and his work with the Kettering Group.

In the early 1990s on a trip to the US, Geoff was invited to speak to the National Security Agency (NSA), the US' classified electronic intelligence gathering and cryptology organization. He probably delivered the latest version of his 'Listening to Satellites' talk. But he also had some fun with the NSA crowd. He introduced himself by saying, 'I'm Geoff Perry, from the KGB.' After a short pause, he clarified, 'The Kettering Group at Bude.'[337] Luckily, the audience found the remark funny.

Max White later recalled that Geoff said, '[The agencies] wouldn't do a question-and-answer session, primarily because if they asked the questions, it would show that they didn't know the answers. He said, "I always knew when we hit home when I was talking about [interesting] stuff." He said, "The pencils would come out and write notes and go back into the pocket again." And he thought, "You didn't know that." We had a pretty good idea of what they knew, and [what] they didn't know, by what they didn't say and how they reacted.'[338]

Having worked with the Stockholm International Peace Research Institute (SIPRI) for several years providing data for their publications, Geoff was invited to their 1981 symposium in Stockholm titled 'Outer Space: A New Dimension of the Arms Race'. At the conference, '20 eminent scientists, lawyers and diplomats from 12 different countries discussed the consequences of the militarisation of outer space as well as further arms control and disarmament measures.'[339] Geoff was in the company of experts from the Brookings Institute, the Federal Aviation Administration, and the Institute for Theoretical Physics, among others. He gave a presentation titled 'Identification of military components within the Soviet space programme,' summarizing his accumulated knowledge from more than 20 years of tracking Soviet satellites. Geoff discussed the latest details around Soviet photographic reconnaissance platforms, electronic intelligence gathering spacecraft, early warning satellites, ocean surveillance and navigation systems, communication missions and Soviet anti-satellite capability.

The US Congress' Office of Technology Assessment was tasked in the early 1980s with evaluating civilian space stations including the Soviet Union's efforts in low Earth orbit, especially the Salyut series of space stations. The OTA in 1982 convened experts for a workshop in Washington, D.C. on the Soviet space programme, which included Geoff. He was joined by such notables as James Oberg, Nicholas Johnson, Craig Covault and Charles Sheldon's colleague and Group member Marcia Smith. Geoff presented a paper titled 'Soviet Space Stations: Achievements, Trends and Outlooks', which was published as a dedicated appendix to the report generated by the workshop a year later. He described how the Soviets used the Salyut platforms and the capabilities, advantages and shortcomings of the stations. The report was reviewed by former NASA associate administrator George Mueller and astronomer Carl Sagan, among others. Although the Soviets had provided a surprising amount of data for the report, there was still enough Soviet-style secrecy in place to keep Geoff and his Group members busy for many more years.

Shortly after his retirement from school in 1984, Geoff attended an international meeting in Sweden, the 34th Pugwash Conference on Science and World Affairs. The 1984 meeting gathered 135 scientists, scholars and public figures to discuss 'rising tensions in international relations and seemingly poor prospects for addressing the worldwide build-up of nuclear and conventional weapons.'[340] Geoff gave a talk on the 'Extent of Militarisation of Outer Space', an updated version of his 1982 SIPRI presentation. Among the many luminaries also speaking at the meeting was Hannes Alfven, winner of the 1970 Nobel Prize in Physics. Geoff's paper was included in the proceedings of the conference and in a separate volume summarizing all the Pugwash Conferences activities during 1984.

The Pugwash Conferences

The Pugwash Conferences on Science and World Affairs focus on nuclear disarmament and ways to reduce conflict and war around the world. In the 1950s, Albert Einstein and English mathematician Bertrand Russell wrote an appeal to world governments to seek peaceful resolutions in place of war. A Canadian philanthropist offered to host a conference in his hometown of Pugwash, Nova Scotia, in 1957 to further the goals of the appeal. The Pugwash Conferences organization received the Nobel Peace Prize in 1995 'for their efforts to diminish the part played by nuclear arms in international politics and, in the longer run, to eliminate such arms.'[341]

In March 1985, Geoff was invited to help judge a competition for pupils run by Independent Television News. The competition's brief was to design a scientific experiment which would then fly on a Space Shuttle mission. At London's Science Museum, four girls from an independent school in Kent were announced as the competition's winners by famed author Arthur C. Clarke. They proposed to run an experiment to see how crystals grow in microgravity. Unfortunately, before their experiment could fly, the *Challenger* accident shut down US crewed space missions for nearly three years. Owing to a backlog of other payloads, the Kent pupils' experiment did not fly until 1992 aboard Space Shuttle mission STS-47, after all four girls had graduated university. While one part of the experiment failed to operate, crystals did grow, and their behaviour was successfully analysed.[342] The experiment was displayed at the Science Museum later that year.

In the summer of 1985, Geoff made a trip to Washington, D.C. to work with Marcia Smith on the latest Soviet space programme report. With Charles Sheldon's passing four years previously, '[Geoff's] role in that final series was much more substantial than the earlier volumes,' said Marcia Smith. He also arranged a visit to Cheyenne Mountain in Colorado, Johnson Space Center in Houston and the Kennedy Space Center in Florida. Geoff's hosts were Nicholas Johnson, Mark Severance and Richard Flagg, respectively. He had the opportunity to watch the launch of Space Shuttle *Discovery* launch (mission STS-51I), which he called 'the fulfilment of a lifelong ambition.' He described the launch in his diary.

> At T-3 seconds we saw a glow around the base of the Shuttle stack which turned into a brilliant white light as it lifted off and cleared the tower. It disappeared quickly into the clouds but reappeared as a truly magnificent sight as it carried on upwards into the upper cloud layer. The sound reached us about seven seconds after lift-off and increased several seconds later when the noise of the [solid rocket boosters] joined in. I could feel the vibration and I stood with my mouth open to equalise the pressure in my ears.[343]

Geoff attended the Radio Amateur Satellite Corporation (AMSAT-NA) Convention in Atlanta, Georgia, USA in November of 1988, and delivered the after-banquet address on 'Listening To Satellites'. Several attendees later fondly remembered Geoff. College professor and amateur radio enthusiast Martin Davidoff recalled sitting with Geoff and

Symposium honoured guest Leonard Labutin, then the head of the AMSAT organization in Russia: 'Hearing Perry and Leo Labutin, [amateur radio callsign] UA3CR, swapping stories after lunch at the 1988 AMSAT Annual Space Symposium was an experience this author will never forget.'[344] German ham radio operator Rolf Neifind attended a separate meeting of AMSAT-UK and remembered Geoff as 'the teacher from Kettering [with] a lot of British humour.' Geoff also attended the 1995 AMSAT-UK Colloquium and won a prize for the best presentation, an extensive analysis of Soviet satellite operations.[345]

AMSAT

AMSAT, or the Radio Amateur Satellite Corporation, is a nonprofit, worldwide, volunteer-driven organization that emphasises education in building and operating amateur radio satellites. The first amateur radio satellite, OSCAR 1, was placed into orbit with the launch of Discoverer 36, one of the early CORONA photo-reconnaissance spacecraft in 1961. Since then, several dozen ham radio satellites have flown in space. AMSAT organizations currently exist in over 20 countries around the world.

In a highlight of Geoff's career, he spoke at the Royal Institution (RI) in London in November 1989. The RI, founded in 1799, promotes scientific education for adults and children. Eminent (and self-educated) British physicist and chemist Michael Faraday started the Friday evening lecture series, known as the Friday Evening Discourse, at the RI in 1825 that still takes place today. Geoff gave his Friday Evening Discourse using a glass rod that Faraday himself had employed nearly a century and half previously to discover the physical principle now known as the Faraday effect. During the talk, Geoff used a laser and an electromagnet to show how the polarization of light is altered by a magnetic field; Faraday had done the same experiment (using an oil lamp) and proved that light and magnetism were related.

During the lecture, Geoff also described the Doppler effect, satellite spin rate decay, the discovery of Plesetsk, decoding Soviet spacecraft telemetry signals, telemetry from the first two Chinese satellites, biomedical telemetry, and military uses of geodetic (measuring the shape and other properties of the Earth) spacecraft. Isabel remembered:

> Right at the end, he concludes and thanks colleagues in the Kettering Group who freely shared results. 'I would not have been able to devote so much of my time and energy to this work without the whole-hearted support of my wife and daughter to whom I offer my heartfelt thanks.' I think that was very important, to have a platform to say that. That was a very special lecture. He was chuffed; all the great scientists of the empire had spoken there over the years. It was a part of history. To be invited to speak there was no small thing.[346]

An expanded version of Geoff's talk was published in the *Proceedings of the Royal Institution* the following year.

Geoff frequently went to the major airshows in Europe, especially those at Farnborough and Paris. Max White recalled:

> The two things that you used to take to an airshow were rolls of film—no digital cameras back then—and the biggest tape measure you could get your hands on. We would stay there hours, every day. The Russian guys are used to this. They used to stand against the wall watching us. There was a new satellite that had gone up called Cosmos 1500. You have the payload, and you have four wings coming out. It was up on a pole, and you wanted to climb up the pole and start measuring, but we thought we'd get thrown out. This was a full-scale model. If you got the length of the antenna, you could work out the frequency, of course. There was one time when we were doing Cosmos 1500, this Russian guy came over, he said, when you finish that, why don't you go to the one over there? We looked over, and [there] was the world's biggest propeller blade. The Russians were laughing, we were laughing.[347]

In the summer of 1990, Geoff taught at a residential Summer

Space School, hosted by Brunel University near London, for 15- to 18-year-olds from the member states of the European Space Agency with a strong interest in space. The Space School provided a week of classes and activities for pupils wanting to learn more about the space industry. In one of those activities, 'Students have the chance to listen to and interpret signals from satellites passing overhead under the watchful eye of Geoffrey Perry in the Tracking Station.'[348] The attendees also conducted computer-based design work, built and launched model rockets and toured space research facilities. The Space School spawned several follow-on London-area pupil space events in the early 1990s.

Geoff was invited to the First European Conference on Space Debris held in Darmstadt, Germany, in April 1993, sponsored by the European Space Agency, the UK's National Space Centre, and the governmental space organizations of France, Italy, and Germany. European space experts were becoming concerned in the late 1980s and early 1990s about the large numbers of inert objects in orbit that had an increasing collision potential with functioning satellites. Over 250 experts from 17 countries attended the conference and presented dozens of technical and legal issues related to the problem of debris.

Geoff delivered a paper at the conference with the verbose title, 'A Possible Mission for the Sub-set of Cosmos Satellites Which Produce Additional Objects During Their Active Phase.' He described several Cosmos spacecraft that had ejected smaller objects, apparently intentionally, while in orbit. He traced the ground track of two such missions and demonstrated that they overflew the Moscow area shortly after releasing one or more small trackable sub-satellites. Geoff postulated that the intent was to provide training for Moscow area anti-ballistic missile system radars, with the small objects simulating multiple nuclear warheads released from a rocket nosecone. He did concede that, with the recent dissolution of the Soviet Union, no such missions had flown since, which he thought might mean that Russian authorities no longer felt them necessary.

As Geoff's renown spread, he received many letters requesting information and asking for his time, typically as a speaker at an event. Before he was famous, Geoff used primarily mail to communicate with colleagues. As facsimile machines became prevalent, Geoff found them more cost-effective than phone calls and lengthy letters. But he continued answering every personal letter sent to him, even with the advent of email in the last ten years of his life. For many years in the 60s and 70s, he traded notes continuously with Desmond King-Hele regarding satellite observations.

Over his entire life, there was a constant stream of requests for him to appear at an array of different organizations to talk about satellite tracking. Geoff exchanged letters with NASA Administrator Tom Paine, several Apollo astronauts, senior Soviet space programme manager Boris Petrov, and other luminaries.[349] He received so many queries asking how to track satellites that he apologetically resorted to a response which, following a brief personalised paragraph, referenced his chapter in the book *Artificial Satellite Observing*, edited by Howard Miles. Geoff gave talks to organizations as diverse as the Institution of Electrical Engineers and various universities, to local groups and a primary school class in the west of England. Such was the demand on his time that he had to politely turn down many invitations to lecture.

15

Author, Contributor, Editor

1967 – 1996

Geoff specialised in sharing his information and findings as widely and as often as he could. He typically wrote with a specific readership in mind, but he could also address a general audience. Below are major book contributions, along with selected articles and papers that represent the breadth and depth of Geoff's knowledge. While he never wrote a book of his own, he contributed extensively to numerous volumes throughout his career. When asked why he hadn't written a memoir, Geoff said he 'just didn't have the time.'

In late 1967, Charles Sheldon completed his first unclassified analysis of the Soviet space programme for the US House of Representatives. In the report, Sheldon wrote, 'The honor [of revealing Plesetsk] was left to some British school boys at the Kettering Grammar School north of London. Their excellent program of tracking uses relatively simple equipment and has been crowned with many successes.'[350] When it was published, the report was the single most informative open-source description of the USSR's space exploration efforts, and according to the *New York Times*, 'read at times like a Cold War spy novel.'[351]

Sheldon credited Geoff and the Kettering pupils for the reveal of Plesetsk and the announcement of it in the press. Describing the Kettering activities, Sheldon wrote, 'By listening to radio signals from spacecraft, they not only find their orbits but even have come to recognise when Soviet recoveries are in process and what beacon signals come from the landed capsule until ground parties reach the payload.'[352] Elsewhere in his report, Sheldon further described the pupils' radio detective work and the hardware they used. *Aviation Week and Space Technology* covered the highlights of Sheldon's report in a three-page lead article in their 'Space Technology' section, including a comment about a 'poorly equipped' Kettering Grammar School Group revealing the location of Plesetsk.[353]

Geoff kicked off a series of articles in *Flight International* in 1968, describing the first 250 Cosmos satellites launched into space by the Soviet Union. The series stretched into September of the following year. The articles were a thorough description and analysis of the various missions and capabilities of Cosmos spacecraft, as best as was possible given the Soviets' fondness for secrecy. The first article focused on the eight-day-mission recoverable satellites launched from Baikonur and Plesetsk. The second report the following May described the Soviets' lunar and interplanetary craft, human-mission precursor flights, FOBS vehicles and meteorology satellites. Finally, in two parts during September, Geoff analysed miscellaneous missions from Plesetsk and spacecraft launched from Kapustin Yar. By 1968, he had become a recognised expert on the objectives, functions and significance of the Soviets' busy space programme.

Charles Sheldon was tasked by the US Senate with compiling a detailed description of Soviet space activities. In December of 1971, he sent a 625-page volume, *Soviet Space Programs, 1966–1970*, to the US Senate Committee on Aeronautical and Space Sciences, which became known as the 'green book' for its light green cover. (This report was separate from the 1967 publication noted previously, which was a relatively brief summary and comparison of the Soviet and US space programmes.) Geoff and many Kettering Group members contributed to the report and were thanked by name. In describing the analysis and findings of Geoff and his pupils, Sheldon wrote in the report:

> With the passage of time, these Kettering techniques have given a highly professional, consistently positive identification to many aspects of the Soviet space program from completely unclassified private sources, which are not matched by any public release of data by the Soviet Union or United States governments. The gen-

> eral public interested in data on Soviet flights owes a debt to Kettering School whose public findings in a few instances have been factors toward influencing the official bureaucracies to ease up on the rigid suppression of what is essentially non-sensitive information about spaceflights.[354]

In the July 1973 edition of *Physics Education*, Geoff published a survey article on the Soviet and American lunar exploration programmes. The emphasis in the story was on the technical and scientific aspects of the two programmes, highlighting their similarities and differences. Geoff showed how, in the first years of the space age, both programmes had similar goals and used similar technologies. Failures were common early on, but both countries' space efforts became more reliable and sophisticated over time. The Americans and Soviets diverged in the late 1960s, however, with the Americans flying crewed Moon missions aided by supporting satellite systems, and the Soviets concentrating on robotic lunar exploration. A footnote at the end of the article invited interested readers to write to Geoff to receive nine tables of data listing the Soviet and US lunar programmes and missions.

Howard Miles, director of the Artificial Satellite section of the British Astronomical Association, was well acquainted with Geoff's work, and invited him to contribute a chapter to Miles' 1974 book *Artificial Satellite Observing and its Applications*. Geoff contributed a chapter entitled 'Radio Tracking of Satellites', describing the methods of monitoring, recording and analysing signals on the shortwave radios in his laboratory. This is one of the most far-ranging pieces Geoff ever composed, from describing construction details for dipole antennae to architecture of Soviet research ships serving as satellite signal reception and relay platforms. He detailed state-of-the-art tracking equipment and procedures at amateur and professional levels. Geoff also explained the Kettering Group's method of using the Doppler shift to determine a satellite's closest approach distance and described how the Faraday effect caused periodic fading of spacecraft signals. Desmond King-Hele and Patrick Moore also contributed chapters to the book.

In a *New Scientist* magazine review of the book, Kenneth Gatland, editor of the journal *Spaceflight* and the president of the British Interplanetary Society, devoted three paragraphs to Geoff's contribution, noting that Geoff had 'succeeded in some quite remarkable discoveries,' such as the failed docking of Soyuz 15 with the Salyut 3 space station.[355] He described Geoff recalling how he was listening to Salyut 3 transmitting at 4:30 in the morning, then looking out his window to see the recently launched Soyuz 15 capsule transiting 17 seconds ahead of the space station. When someone asked Geoff how they could learn about satellite tracking, Isabel remembered, *Artificial Satellite Observing* 'was always the book he would recommend.'[356]

In the summer of 1976, the latest version of the 'green book' covering the years 1971–1975 was published by the US Senate. Sheldon had invited Geoff to write sections of the 880-page report, complete with by-line, on the Soviet reconnaissance satellite programme, their navigation spacecraft constellation, and the Molniya communication system. In the report, Sheldon wrote:

> Analysis of [recoverable Cosmos] missions has taken much time, which is commensurate with their importance. The world leader in such work, as far as the public domain is concerned, is Geoffrey E. Perry, whose name has come up in many other contexts in this study. Because his original international reputation was built on his ability to understand the variations in these Cosmos flights, he has been invited to make some direct contributions to this study, and two of them are offered as annexes to this study (as well as an additional study as an annex to a different chapter)....The continuing work over lunch hours provided unconscious training in the 'scientific method' to successive generations of pupils, some of whom are named in the annexes to this chapter for their specific contributions.[357]

Aside from his writing 100+ pages, Geoff is named or cited in Sheldon's report no fewer than 41 times. Kettering Group members also named in the report include Philip Perkins (Molniya), Sven Grahn (navigation satellites), Ian Wildman (navigation satellites), Paul Rosser (human mission telemetry), and David Dean (human mission voice communication).[358] Dean's observations, combined with other information, led Geoff to conclude at the end of the Human Mission Report segment that the Soviets were conducting two flight programmes in their Salyut space station system: a civilian project flying in higher orbits, and a classified military programme at lower orbital heights. In fact, Salyuts 3 and 5 were staffed by Soviet military officers as part of the then-secret Almaz reconnaissance project. Almaz was cancelled after three missions—only Salyut 5 was considered even moderately successful—as robotic reconnaissance satellites proved more effective.

Charles Sheldon contributed to another sizable report for the US Congress in 1977, this one titled *World-Wide Space Activities*. Sheldon wrote several sections of the report, including a description of the nascent space programme of China. The volume's introduction thanks 'Geoffrey E. Perry, leader of the Kettering Group, the United Kingdom, for his help as a consultant and reviewer.'[359] Geoff, Robert Christy, a Swedish member of the Group, Jan-Ola Dahlberg, and Richard Flagg are acknowledged by Sheldon for their information and assistance with his chapter on China's space programme.

Also in 1977, Geoff wrote a series of articles for the British publication *Royal Air Forces Quarterly*, a journal of equipment, history and policy articles related to the RAF whose authors were almost exclusively military. Alongside senior RAF officer autobiography excerpts, aircraft testing notes and air warfare strategy discussions, Geoff described the various Soviet satellites systems he had been studying and writing about for many years:

- Summer 1977: 'The Molniya Communications Satellites'
- Autumn 1977: 'Russian Geostationary and Early Warning Satellites'
- Winter 1977: 'Russian Hunter-Killer Satellites Experiments'
- Spring 1978: 'Russian Ocean Surveillance Satellites'
- Autumn 1978: 'Soviet Navigation Satellites'

All the articles were oriented to a military audience and updated with the latest information and analyses.

Competition in space between the two superpowers was getting more attention around the world. SIPRI, the Swedish think tank, issued a volume about military satellite efforts called *Outer Space—Battlefield of the Future?* The book described various military space-based systems used by the US, the Soviets, and other countries. The volume's author credits the Kettering Group throughout the text for providing valuable information and cites many of Geoff's published articles and private communications as authoritative data sources. The Institute has published an annual yearbook since 1968, detailing the state of the world's military systems and efforts to restrain and regulate arms. The yearbooks often relied on Geoff's publications and advice in the 1970s and 1980s.

Before his death in 1981, Charles Sheldon had worked for several years on his next major update on the Soviet space programme for the US Congress, *Soviet Space Programs: 1976–80*. The report, which was completed by his colleague Marcia Smith and other Congressional Research Service experts after Sheldon's death, spanned three volumes. The first volume was released in late 1982; the book's cover letter to Congress, written by then-Senator and *Apollo 17* Moon-walker Harrison Schmitt, acknowledged Sheldon as 'one of the free world's foremost authorities on Soviet space activities.'[360] Geoff was credited with rewriting and revising the first chapter on Soviet space facilities and launch vehicles and updating related data tables. Sven Grahn was also thanked for his peer review of a portion of the text. The second and third volumes would be published over the next few years.

Geoff updated his *Royal Air Forces Quarterly* piece on geostationary and early warning satellites. He had new information about Soviet early warning spacecraft for the article and included more graphics showing potential viewing angles for the warning platforms. Geoff presented his revised article during the BIS' Military Satellites Programmes seminar in January 1981 at the Society's headquarters in London. He later published the paper in the February 1982 edition of the *Journal of the British Interplanetary Society*. The BIS talks were typically open to the public, and that night there was probably one special attendee. According to Max White, 'Every time he'd do a lecture in London there'd be a representative from the Russian embassy.'[361] The Soviets were undoubtedly curious as to what Geoff was saying about them.

Charles Sheldon's office, the Congressional Research Service, compiled information in the 1970s on US and Soviet space travellers titled *Astronauts and Cosmonauts Biographical and Statistical Data.* Sheldon's colleague, Marcia Smith, assembled the first volume and assisted with periodic updates. This volume is a reference resource, as opposed to the 'green books' that described the Soviet space programme overall. The 1985 revision included inputs from several sources including Geoff.

Soviet space programme expert Nicholas Johnson compiled many authoritative texts on what was known in the West about the Soviet Union's space operations. He published a series of books from 1984 to 1990 that described all significant Soviet space activities from the previous year. His 1985 volume noted that he derived information for the book from 'highly respected sources such as… the English Kettering Group, headed by Geoffrey Perry.'[362] One of that edition's appendices is a table of the radio frequencies and telemetry types received from Soviet satellites over the previous year, all the data for which was supplied by the Kettering Group. Geoff's articles and information provided are cited in the book's footnotes.

Johnson compiled the first five volumes into a publication for the American Astronautical Society, to which Geoff also contributed. As a consultant for the US Air Force's Phillips Laboratory in the early 1990s, Johnson assembled a study for the US Air Force called *Europe and Asia In Space*, with two editions published between 1992 and 1994, updating his 1980s reports on the Soviet space programme and including the rest of the world's space efforts. Geoff's work is referenced and cited several times, and the first appendix, a list of foreign satellites' transmission frequencies, was assembled by the Kettering Group. In the acknowledgements section of the book, both Geoff and Sven Grahn are thanked for providing 'technical information and other assistance of significant value.'[363]

In 1985, the American Radio Relay League, the principal ham radio organization in the US, published *The Satellite Experimenter's Handbook* by physicist and amateur radio expert Martin Davidoff. Along with detailed information about amateur radio operators' utilization and operation of satellites dedicated to all modes of ham radio communications, Davidoff included a section called 'Satellite Sleuthing' referencing Geoff and the Kettering Group. Davidoff commented, 'The Kettering Group has on occasion, for example, left the professionals dumbfounded by decoding downlink telemetry without prior knowledge of the content or format.'[364] Davidoff also mentioned Kettering Group member Greg Roberts of South Africa, cited an article by Geoff and Chris Wood, and described an analysis involving Geoff's colleague Cliff Ranft. He encouraged readers to do some satellite tracking of their own, complete with recommended methods. He also recounted how he determined the exact launch time of Sputnik I with information supplied by Geoff that Davidoff considered the final word.

Desmond King-Hele asked Geoff in 1987 if he would become an editor and consultant for the authoritative volume *The RAE Table of Earth Satellites.* Geoff had relied on the book's data for many years. He compiled information for the monthly update of the book and helped edit several editions of the main volume; the last version appeared in 1992. Geoff also co-edited and wrote a chapter for the 1993 book, *DRA Table of Space Vehicles*, describing all the world's satellites that, to that time, had been launched beyond Earth orbit. The UK's Defence Research Agency (DRA) had absorbed the Royal Aircraft Establishment in 1991. The DRA maintained a version of *The RAE Table of Earth Satellites* with Geoff's help until 1994. It was succeeded by *The Kettering Tables of Earth Satellites*, which was compiled by Geoff, first published in 1993 and updated several times in the 1990s.[365]

In the mid to late 1980s, Soviet leader Mikhail Gorbachev pioneered what he termed 'perestroika' (restructuring) and 'glasnost' (openness). Geoff wrote a thoughtful essay for *Space Policy* journal in 1989 on how the Soviets' reforms had 'affected their space programmes, their relationships with other countries in space matters and the ways in which their media dealt with space events.'[366] Geoff contrasted the official disinformation around the failed 1971 Soyuz 15-Salyut 3 docking with the recent, albeit slow, progress in revealing secret launch sites, classified rockets, and disguised satellites. He noted how the Soviet Union had previously denied any failures of their space efforts but had lately started admitting mistakes and conducting some self-criticism.

In the piece, Geoff also described the recent commercialisation programme via the formation of a Soviet civilian space agency that was advertising launch and satellite services. He commended the relaxation of traditional secrecy around Soviet space activities even though more openness had 'removed some of the challenge and excitement which has fuelled the Kettering Group and maintained their interest for over thirty years,' but insisted that the Group would 'continue to pursue their hobby of keeping watch on the Soviet space programme (and the closed programmes of other nations).'[367]

Geoff pursued his long-standing consulting work with the US' Congressional Research Service even after the passing of Charles Sheldon. As mentioned previously, Sheldon's colleague Marcia Smith continued his research and analysis for the US Congress and in May 1988 the first section of the last version of the 'green book', *Soviet Space Programs: 1981–87, Part 1,* was published. Geoff wrote a major section of the report on launch vehicles, launch sites and tracking support. He also prepared several tables of data included in the volume. A number of Geoff's previously published articles were cited as sources of information. Kettering Group member Max White wrote the appendix on Soviet tracking ships in the space support role.

Later that year, Smith released a summary report for the US Congress on *Space Activities of the US, Soviet Union and Other Launching Countries, 1957–1987*, one of several in a series that started in 1982; Geoff contributed to all editions. She thanked Geoff as the only named reviewer of the report. In early 1989, Part II of Smith's 1981–87 report was published, with Geoff writing chapters on space applications and military space activities. Richard Flagg wrote an appendix for the report on Soviet lunar exploration, assisted by Geoff, Sven Grahn and Mark Severance. Geoff also contributed many of the report's data tables and had numerous articles cited. Smith said, 'The last two [green books] of the series, 1976–1980 and 1981–1987, were written after

Charles died and I can state authoritatively that they could not have been completed without Geoff and the Kettering Group.'[368]

Former *Daily Mirror* science correspondent Arthur Smith asked Geoff to write the foreword for his book *Planetary Exploration* in 1988. In the foreword, Geoff reflected, 'I consider myself privileged to have lived through the exciting early years and growth to maturity of the space age.'[369] Geoff described the thrill of touring behind the scenes at the Jodrell Bank Observatory and waiting breathlessly at NASA Headquarters for the first images from the Viking 1 Mars lander. Later that year, he co-edited a book about the exploration of Mars with the ITN space reporting team headed by his colleague at the network, Frank Miles. The colourful and detailed book included background on both the US and Soviet efforts to reach Mars with interplanetary probes, Mars mission profiles, human flight considerations and a lucid explanation of orbital mechanics.

Geoff wrote a chapter called 'The Russian Satellites' for the 1990 book *Weather Satellites* by the American Meteorological Society. He detailed the development and operation of the Soviets' Meteor meteorological satellite system. Geoff also described the early efforts at a Soviet geostationary weather satellite system, GOMS. He concluded with a short section on oceanographic Cosmos satellites. Included in the article were data from Kettering Group members Les Currington and Peter Wakelin.

Desmond King-Hele, Geoff's long-time colleague, wrote a career memoir in 1992, *A Tapestry of Orbits*. In the book, King-Hele recounted his professional experiences at RAE Farnborough over four decades, detailing his extensive work in jet engine development and satellite-tracking-derived science. As we've seen, his activities in visual satellite observation and Doppler effect orbital determination were inspirations to Geoff. He mentioned Geoff several times in the book, particularly regarding Geoff's work on *The RAE Table of Earth Satellites*. Also, wrote King-Hele, 'Geoffrey Perry told me that [King-Hele's first book] *Satellites and Scientific Research* aroused his interest in space, which led to his many years of assiduous tracking and analysis of the USSR satellites from the Kettering school where he taught physics.'[370] However, by the time *Satellites and Scientific Research* was published in 1962, Geoff had already been observing and listening to satellites. In their 1962 article about conducting HF radio satellite observations, Geoff and Derek cited John Osborne's 1957 *Short Wave Magazine* article and G. Boudouris' 1960 paper in the *Journal of the British Institution of Radio Engineers* as sources for Doppler tracking information and methods.

Geoff continued his association with Marcia Smith. Smith published regular updates on the Soviet, then the CIS (Commonwealth of Independent States)/Russian space programme for the US Congress throughout the 1980s and 1990s. Smith wrote a report on the Chinese space programme in 1998, and the only external collaborator she credited with contributions to the report was Geoff.[371] He also wrote a monthly column, 'What's Up', under the whimsical pseudonym, 'Perigee'—a play on Perry, G.—for the National Space Society's *Ad Astra* magazine in the mid-1990s. The column was a catalogue of the latest satellite launches and spaceflight events from around the world.

16

Influence on Education

1966 – 1990

Geoff didn't set out to change how science was taught in Britain. But he certainly made a difference in the lives of pupils and teachers, in Kettering and eventually across the UK and the world.

Being a regular reader of the journal *Physics Education*, Geoff noted a question previously posed in a 1966 edition of the magazine. The journal's editor had wondered, 'Is there a way to use radio signals from satellites in teaching astronomy?' Geoff apparently had a visceral reaction to this question and composed an opinionated letter that appeared in a later edition of the magazine, which began: 'My immediate response to this question is to answer quite definitely, No!'[372] Aside from using satellite signals to help verify Kepler's Laws (which describe the motion of objects in orbit), he suggested that the physics curriculum could be enriched through observations of the Doppler effect and other phenomena revealed by spacecraft transmissions. He also noted that visual observations of satellites promoted familiarity with the night sky, so using both the radio signals and direct observations could engage pupils and leverage their enthusiasm.

Also in *Physics Education*, Geoff commented on an editorial that had appeared in an earlier issue. The journal had published an opinion piece about the state of mathematics teaching in Britain. The editorial writer had decried the breadth and depth of mathematics courses as they failed to provide adequate support for the requirements of the physics courses at that level. Geoff generally concurred with the comments and suggested that, if mathematics courses did not support the sciences being taught at schools, then the sciences would have to teach the necessary mathematics themselves.

Later, *Physics Education* marked its 21st anniversary in 1987 with a look back at the intervening years and the change—or lack thereof in some cases—in how physics was being taught and how teachers were faring. One of the milestones noted during the first year of the Journal's existence was that 'Geoffrey Perry was already determining the orbital periods of satellites from measurements he was making at his then Grammar School in Kettering.' [373] The story also lauded John Osborne, author of one of the first published articles on Doppler analysis of satellite transmissions, whom Geoff had cited in his early writing.

In 1972, John Percy, professor of astronomy at the University of Toronto, visited Geoff when Percy was in England. While on sabbatical at the University of Cambridge, he contacted Geoff to learn more about the Kettering Group's activities. Geoff invited him to come to Kettering and, among other things, be the audience for the rehearsal of an upcoming talk Geoff would be giving at the University of Oxford. Calling the achievements of the Kettering Group 'so unique and so spectacular', Percy also encountered other creative teaching projects and methods at Kettering Grammar School: an archaeology programme in which students were helping to excavate the ruins of a nearby medieval village, and a pupil weather-monitoring operation that had assisted in a local court case. In a science teachers association newsletter, he wrote about Geoff, 'His purpose was, and still is, to make science more real to his students.' Percy encouraged other educators, writing, 'The number of useful research projects for schools is probably limited only by the ingenuity and imagination of the teachers, students and professional scientists concerned. What could you do in *your* department?'[374]

Geoff shared stories of his interactions with international partners in the *New Era* educational journal. Along with describing his investigation into Cosmos 482, Geoff also wrote about how the Stockholm International Peace Research Institute (SIPRI) asked him to help with a study of the world's reconnaissance satellites. In addition, he related how his 'distant cousin' in Fiji, Chris Wood, had helped confirm that many Cosmos short-mission satellites transmitted their recovery beacon throughout the spacecrafts' final orbit.

The Kettering Group's activities, publicised at length by Geoff and Derek, appear to have had a noticeable impact on how physics was taught in UK secondary schools by the early 1970s. In a 1974 *Spaceflight* article, a British educator described how the physics classroom was much changed from previous years:

> Curriculum changes have brought science, and physics in particular, more into line with the world outside the classroom, while the emphasis has shifted from teaching to learning. Pure blackboard and lecture demonstration methods have largely given way to pupil-based activities either individually or in small groups using an impressive array of modern apparatus....So now it is not uncommon to see basic courses in electronics introduced in the first and second years, i.e., with 11- to 13-year-old children while higher up in the school their colleagues often reach a remarkable level of ability and are encouraged to design and build interesting projects based on their earlier mastery of fundamentals. Their physics lessons are no longer complete in themselves because the children are actively encouraged to apply this knowledge to problems they will meet in the world of technology outside the classroom....A recent article in this journal by a distinguished colleague well illustrates the sort of advanced projects which are being successfully undertaken by schools up and down the country.[375]

The 'distinguished colleague' referred to is Geoff, who had written the cited article about the latest analysis of Cosmos spacecraft by the Kettering pupils. The improving methods of physics education described in the article were certainly a goal of Geoff's work and extensive writing activities.

In early 1978, *New Scientist* magazine commented on the lack of programmes for European pupils to become involved in space research.

The author noted that US high school students had run experiments on the Skylab space station in the early 1970s. Apparently, the European Space Agency favoured getting pupils involved in the upcoming Spacelab missions that the new Space Shuttle would take into orbit. The commenter, encouraging these efforts, closed with: 'ESA should publicly declare its intention to involve schools in space as soon as possible. There is a new generation of high-school students, such as Geoff Perry's world-famous satellite tracking lads at Kettering, who would not waste such an opportunity.'[376]

Geoff's influence on science education continued to reverberate into the 1990s. H. Paul Schuch, professor at the Pennsylvania College of Technology, used Doppler effect measurements in his classroom to teach students how to conduct observations and determine satellite characteristics. Schuch acknowledged Geoff and his pupils' efforts in satellite analysis: 'This avocation has been raised to the level of high artform by Geoff Perry and others of the legendary Kettering Group in England, and the techniques discussed here should certainly be attributed to them.' Schuch concluded, 'As an instructional tool, the satellite may rank in importance second only to the chalkboard.'[377]

A plea for more advanced teaching of Earth science in the journal *Space Policy* also cited Geoff's work motivating his pupils. 'Geoffrey Perry, that ubiquitous and constantly curious educator required [his pupils] to work through the physics and mathematics for plotting launches and orbits.'[378] The author advocated advanced technology in the classroom, better data visualisation tools, increased communication between scientists, teachers and pupils, and a strong relationship between government, industry and education—all ideas with which Geoff would certainly have agreed.

17

Honours and Awards

1973 – 2000

Geoff said, 'You don't expect to get MBEs for pursuing your hobby, do you?' He regarded any recognition he received to be for the entire Kettering Group, not just for him.

The population of the United Kingdom in the early 1970s was about 56 million. Of that number, 325 were appointed as a Member of the Order of the British Empire (MBE), in the 1973 New Year Honours List. The MBE and other awards are given each year to the most prominent and accomplished UK citizens. In particular, the MBE recognises 'achievement or service in and to the community which is outstanding in its field and has delivered sustained and real impact which stands out as an example to others.' In 1973, one name on the MBE List Civil Division was Geoffrey Edward Perry, for being the 'Founder and Leader, Satellite Tracking Group, Kettering Grammar School.' He was in good company; well-known Britons such as the Beatles had previously been awarded MBEs.

Geoff received his award in person from HM Queen Elizabeth II at a Buckingham Palace investiture on March 13. Geoff wrote later about the ceremony: 'Her Majesty asked, "What is your field?" Somewhat taken aback, as I had not expected more than a few polite words, I said, "Satellite tracking." She looked puzzled and put her head to one side and said, "Where?" I replied, "Kettering Grammar School." Thereupon she smiled and said, "Oh, I've seen you on television. How long has this been going on?" I answered, "Since 1961, Ma'am." She then said, "You must find it very exciting," and, for want of something more intelligent to say, I grinned and said, "Well, it keeps me out of mischief.'"[379]

Geoff's MBE decoration.

In March 1974, Geoff received another honour, barely a year after his MBE, when the UK's Royal Astronomical Society presented him with its Jackson-Gwilt Medal, the Society's second oldest medal. Since 1897, the Medal has been periodically awarded to an individual 'for the invention, improvement or development of astronomical instrumentation or techniques, for achievement in observational astronomy, or for achievement in research into the history of astronomy.' The Award's citation specified Geoff's 'work on tracking artificial satellites and interpreting their radio signals'—activities not directly related to astronomy but, as Geoff had earlier observed, pupils were learning astronomical information as they conducted visual observations.[380]

The Medal was accompanied by a gift of £50; Geoff later said he had spent this on a new antenna and pre-amplifier for the school's shortwave equipment. The Jackson-Gwilt medal had also been awarded to such luminaries as Clyde Tombaugh, discoverer of Pluto, and renowned British astronomer Patrick Moore—although Moore received his Award three years after Geoff. Moore had no formal training in astronomy, so he was technically, like Geoff and Tombaugh, an amateur scientist.

Jackson-Gwilt Medal of the Royal Astronomical Society.

Just after Geoff's retirement in 1984, the Kettering Group received a major award, this one from the Royal Aero Club (RAeC) of the United Kingdom. The RAeC is one of the oldest aeronautical associations in the world. Created in 1901, it was a founding member of the world's governing sport aviation body, the Fédération Aéronautique Internationale (FAI). In 1976, the President of the RAeC, His Royal Highness The Prince of Wales, presented the first of what was to become an annual award, the Prince of Wales Cup. The award was for the 'the most meritorious performance, feat or event by a team or group during the previous year.' The cup was awarded to the Kettering Group for their notable activities during 1983, especially the Group's work concerning the re-entry of the nuclear-powered Cosmos 1402 satellite. The award citation read 'for the work of Kettering Boys' School on radio monitoring of spacecraft and satellites.'[381]

Geoff and members of the Group received the award in the Aeronautical Gallery at the Science Museum in London in December 1984. Geoff later remembered the ceremony at the Museum as being unique in the history of the Kettering Group: 'There were people who had heard of each other but had never actually met. There were 21 of us there that night. It was the largest and smartest ever gathering, I suppose, of the Group.'[382] Many of the men that evening wore a special necktie featuring the Kettering Group insignia. The woman in attendance—Jean Perry, Isabel Carmichael, Marcia Smith and Sarah Mobbs—each wore a special silver brooch commissioned by Geoff and designed to replicate the Group's logo. Interviews for Patrick Moore's 1985 radio programme about the Kettering Group were recorded that evening after the award ceremony.

The Kettering Group, 1984. Geoff is holding the Prince of Wales Cup. Top row from left: Richard Flagg, Sven Grahn, James Bambridge, Grant Thomson, Michael Cartwright, Max White, Marcia Smith, Sarah Mobbs, Charles Carmichael. Middle row: Robert Christy, David Hawkins, Jean Perry, Geoff Perry, Isabel Carmichael, Peter Wakelin, Derek Slater. Bottom row: Les Currington, Andrew Driver, Mark Severance, Chris Wood.

Prince of Wales Cup

Kettering Group brooch

Geoff was elected to Fellowship of the Institute of Physics, one of the largest physical science societies in the world, in 1968. The IOP had presented awards for various achievements for many decades, and in 1986 the Institute inaugurated the Teachers of Physics Award for 'teachers who have, by their outstanding teaching of physics, inspired others to continue with and enjoy the subject.'[383] Geoff was among five secondary school teachers honoured in the first year of the award. A certificate and a Waterford crystal paperweight were presented to him at the IOP Education Group's Annual Meeting at Christ's College, Cambridge, in March 1986. He sometimes used the post-nominal 'FInstP' in his by-line and signature block: Fellow of the Institute of Physics.

Crystal paperweight accompanying the Teachers of Physics Award from the Institute of Physics.

Geoff continued to be recognised long after his retirement. The following universities, including the University of Reading, his alma mater, conferred Geoff with master's level honorary degrees: Loughborough University (1978, Master of Technology), University of Reading (1980, Master of Philosophy), University of Leicester (1986, Master of Science) and The Open University (1989, Master of University). As he accepted his degree from the Open University, he said:

> It is with a combination of pride and humility that I accept the award of the honorary degree of Master of the Open University. Pride in that my hobby of the past 30 years or so has been recognised by the University as being worthwhile and that my work with the pupils of the Kettering Grammar School has not passed unnoticed. Humility because I realise that implicit in the honour you accord to me is a recognition of the efforts of the many individuals, in this country and around the world, who, by drawing on their expertise and contributing their observations, comments and time, have supported and encouraged me in my study of the various space programmes.[384]

After the Kettering Boys School closed in 1993, a new educational facility, the Tresham Institute, was founded on the site. On 2 September 1999, Geoff had the honour of opening Tresham Institute's science block and naming it 'The Perry Science Centre,' which included three laboratories. At the ceremony, Geoff said, 'Kettering Boys' School used to be on the same site and was always a centre of academic excellence, and I am sure it will be again.'[385] This was the second building named after Geoff, the first being a computing complex at South Devon College near Plymouth, the Perry Data Centre.[386]

Geoff at the dedication of the Perry Science Centre.

Geoff was accorded two final honours posthumously. *Aviation Week and Space Technology* has published an Annual Laureate Awards list since 1957 recognising extraordinary achievements in aviation, defence, and space. In the 1999 Space category, announced in February 2000, the first person honoured was 'Geoffrey E. Perry, acclaimed sleuth of Russian space activities and founder of the Kettering Space Observer Group in England, who died Jan. 18. Perry, an unassuming schoolteacher, turned an educational project for his physics students into a 35-year quest, uncovering the secrets of the Soviet and Russian space program.'[387] Moreover, in March 2000, on the recommendation of Harvard astronomer Jonathan McDowell, a committee of the International Astronomical Union named two asteroids in honour of Geoff and the Kettering Group: 10289 Geoffperry and 10290 Kettering.[388]

18

Legacy

Geoff's remarkable career in education extended across five decades. His satellite tracking efforts and his related work with Kettering Grammar School pupils and adult volunteers around the world spanned some 40 years. He was one of those people who seemed like they would live forever. Geoff's death was sudden and shocking to his family, friends and colleagues.

Perhaps the last mystery Geoff worked on was a series of signals on an unusual frequency reported by an observer in late 1999. Geoff saw the report, received more information from the observer and enlisted Sven Grahn to assist in the investigation. Two different sets of signals appeared on separate VHF frequencies. With all the information he had gathered, Geoff derived the orbital data and plotted trajectories for two suspected satellites. He then searched an online orbital information database to seek a match with listed satellites. The vehicles turned out to be part of a United States classified system called the Naval Ocean Surveillance System (NOSS), which monitors radio and radar signals from sources at sea and determines position and velocity of the transmitters.

In Geoff's last public comments on the Russian space programme, he was cited in the *Aviation Week and Space Technology* 1 January 2000 issue. He confirmed the launches of a Russian ocean surveillance satellite and Cosmos 2358, a missile warning platform.

Less than three weeks later, on 18 January 2000, Geoff Perry, age 72, passed away from a heart attack in Cornwall. All the outlets that had carried Geoff's announcements and discoveries wrote tributes to him.

- *The Times*: 'Geoffrey Perry became one of the world's foremost authorities on the Soviet rocket and satellite programme and put himself and his school on the space exploration map.'[389]
- *BBC*: 'Friends and colleagues have said that they will greatly miss his scientific rigour, his no-nonsense attitude to life and his warm personality.'[390]
- *New York Times*: 'He did amazing things with very little equipment,' said Group member and space analyst Marcia Smith. 'The most important pieces of equipment he had were his brain and his desire to share information with the world.'[391]
- *United Press International*, by James Oberg: 'Perry was affectionately regarded by his colleagues as an archetypical 'English eccentric amateur' who had better knowledge and intuition about his area of expertise—secret space activities—than did academics or even government intelligence officials.'[392]
- *Aviation Week and Space Technology*: 'Perry continued to provide detailed information to *Aviation Week* on Soviet and Russian space activities for 35 years. His analysis of Russian and Soviet capabilities has been used often by the U.S. Library of Congress and other U.S. and European agencies.'[393]
- *Spaceflight* magazine: From Australian spaceflight expert Jos Heyman: 'The night before he passed away, Geoff remarked that he had been lucky as he had been around in the right place at the right time doing the right thing. I think that phrase should be recast to simply read that we were all lucky that we were allowed to be around Geoff.'[394] From Rex Hall, fellow BIS space researcher: 'His talks were informative with him delighting in unveiling his research layer by layer. He always used his sense of fun to illuminate his talks. They were easy to understand, even if very complex, probably a result of being an excellent teacher. I wish he had taught me science. He will be missed by many who heard him, read his work or just learnt about this teacher who reached beyond the classroom.'[395]
- *Wired* magazine: 'He had this quality of self-aware eccentricity,' [Jonathan] McDowell says. 'His importance was that he com-

bined that with the highest standards of research. He showed you could compete with the CIA using the cheapest equipment and a class full of high school students. And in some ways, he was a precursor to the Internet generation…You can do it yourself and establish a reputation based on what you can do on your own.'[396]

- *Jonathan's Space Report*, also by Jonathan McDowell: 'I'm glad that Geoff lived long enough to see the declassification of the former Soviet missions and discover that the "answers in the back of the book" agreed for the most part with his own sleuthing, but I'll miss his insight and comprehensive knowledge.'[397]
- *Space News:* 'Using the tools of amateurs and the skills of professionals, [Geoff and his pupils] revealed just how the Soviet Union was using its new expertise in space. They studied the behavior of spacecraft closely and were able to perceive the small details that formed the bigger picture.'[398]

Geoff's life was celebrated at a memorial service which was held at the Tresham Institute, Kettering, in September 2000 where Geoff had been honoured a year earlier when the science block was named 'The Perry Science Centre.' Along with numerous family and friends, many Kettering Group members attended, and several eulogised Geoff. Fittingly, one reading was from the book that had inspired Geoff as a young man, *The Exploration of Space*, by Arthur C. Clarke. Clarke imagined a future historian looking back at the twentieth century: 'The coming of the rocket brought an end to a million years of isolation. With the landing of the first spaceship on Mars and Venus, the childhood of our race was over and history as we know it began.'[399]

Paul Brand (Jean Perry's brother), David Dean, Derek Slater, Sven Grahn, Marcia Smith and Isabel and Charles Carmichael shared stories and memories of Geoff at the service. Dean remembered the 'tremendous camaraderie in the Group. We regarded ourselves as something special—often to the annoyance of other pupils and sometimes teachers as well.'[400] Derek honoured Geoff by pointing out 'all the people in this room whose lives he touched and enriched and the hundreds of others who are not here.'[401]

Grahn considered Geoff a role model and recalled that he 'created a magic circle around himself. He was a connector, a person with a special gift for bringing people together.'[402] Smith read part of a letter from US Senators John McCain and Bill Frist. '[Geoff's] legendary ability to acquire satellite data using rudimentary equipment and use that data for incisive analysis is one of the many legacies he leaves.... We were grateful to continue to have his wisdom and analysis available to us after he retired from teaching.'[403] Isabel Carmichael concluded the tributes by thanking everyone and giving poignant reflections of her late father's interests and family life. Following this, Charles Carmichael read the citation which appears, together with the Kettering Group logo, in the Book of Remembrance at the Glynn Valley Crematorium, Bodmin, Cornwall:

PERRY, GEOFFREY EDWARD MBE
4th August 1927 – 18th January 2000

FOUNDER OF THE KETTERING GROUP

A man of rare accomplishment who
had the respect of those he
enthused and inspired through his
friendship, teaching and renowned
satellite work.

Remembered with great love and affection

Jonathan McDowell later recalled the weekend of the memorial service:

> We had dinner at the hotel and then there was a service at the school. Then Derek Slater had a party at his house. That's when we got to see Derek's radios that he used for Cosmos interceptions. Apparently that was original equipment from back in the day. It was very moving. It was a sense that they did pretty damn well, now that

> we get to look up the answers in the back of the book with the declassification of all that Russian stuff. They did a remarkable job and really I think set the standard for what we now call open-source intelligence, in a different era.[404]

After Geoff's passing, the Kettering Group formally ceased to exist. However, some former members continued to maintain contact with one another, and in 2002 they organised a Memorial Colloquium, meeting at the Kettering Park Hotel. Papers and updates were delivered, and Derek and Mary Slater hosted a barbeque for attendees at their home in the evening. But, as Sven Grahn said later, 'I think we exhausted the topic, and it didn't happen again after 2002.'[405] And, perhaps, it just wasn't the same without Geoff.

The Kettering Grammar/Boys' School buildings, having housed the Tresham Institute for several years, were demolished in 2007 to make way for a new facility, the Tresham College Kettering Campus; one of the Kettering Group's primary radio antennas was reportedly placed in storage there.

In Kettering's town centre there is a timeline of important local events, which includes this entry: '1966—Russian Satellites: Grammar School Beats NASA'. Though it is well intended, this is not quite accurate; NASA was never tasked with finding secret missile bases, and Western intelligence was aware of Plesetsk long before the first Cosmos spacecraft was launched from the then-secret facility.

Geoff's influence, inspiration and information continue to provide value and insight and is acknowledged by researchers and authors studying the Soviet/Russian space programme. In their book *The First Soviet Cosmonaut Team*, space history researchers Colin Burgess and Rex Hall acknowledged Geoff as a leading space detective. Brian Harvey, a BIS member, described Geoff's efforts to reveal details about the Russian Arkon and the German/Japanese Express programmes in his book *Russia In Space*, and cited Geoff and Sven Grahn as information sources. Rex Hall and David Shayler, in their book about the Soviet Vostok and Voskhod programmes, *The Rocket Men*, acknowledge 'the work of the Kettering Group...and the inspiration of Geoff Perry, who showed us what "space sleuthing" could reveal.'[406] Phillip Clark, a mainstay of the BIS' Soviet/Chinese Technical Forum, was also influenced by Geoff's work.

Many years after leaving school, former pupils of Geoff's still had strong memories of his classroom. One recalled, 'I was lucky enough to have Geoff Perry as my physics teacher in the early 1970s. In addition to being a great physics teacher he was indirectly a great maths teacher. I cannot forget being amazed at his ability to do complex multiplication and long division on the blackboard, whilst looking at the class.'[407] Another remembered, 'I had Geoff Perry as my physics teacher in the late 1950s, and he inspired a reluctant student to achieve a GCE [A-level] in Physics. Later, as a police officer, I met him directing traffic when I arrived at the scene of a road traffic accident, to be greeted with, "Hello, young Hopkins, how long should I do this?" A truly inspiring and modest man.'[408]

The most active of the former members of the Kettering Group in terms of writing about spaceflight and satellite tracking are Sven Grahn and Robert Christy. Both Grahn and Christy run extensive websites about current and past spaceflight activities, including their involvement with the Kettering Group.[409] Grahn published a book about his satellite tracking activities both with the Group and through his independent efforts; he also contributed to a book about Western Soviet space experts. Christy has written several articles about the Kettering Group and other spaceflight topics.

In 2011, Derek Slater gathered his papers and equipment for donation to the National Space Centre in Leicester. In a letter to *The Times*, Derek remembered, 'Together [with Geoff], we worked hard to try to make Kettering excel in turning out science graduates and other citizens with an appreciation of the rigours and joys of science. I hope we succeeded.'[410]

Derek Slater passed away in March 2015. The following year, a permanent exhibit of the Kettering Grammar School Satellite Tracking Group's early radio and recording equipment and accomplishments opened at the National Space Centre. A number of Kettering Group members and family attended the formal opening which culminated in

Isabel Carmichael and Mary Slater cutting a cake replica of the reel-to-reel tape recorder used by the Group to help make their historic discoveries. That day, British astronaut Tim Peake posted on Facebook from the International Space Station, 'Kettering Grammar School Satellite Tracking Group monitored space and satellites for nearly half a century. Celebrating 50 years of pioneering space research as an aid to teaching.'[411]

Mary Slater, left, and Isabel Carmichael cut the replica tape recorder cake at the opening of the Kettering Group exhibit at the National Space Centre.

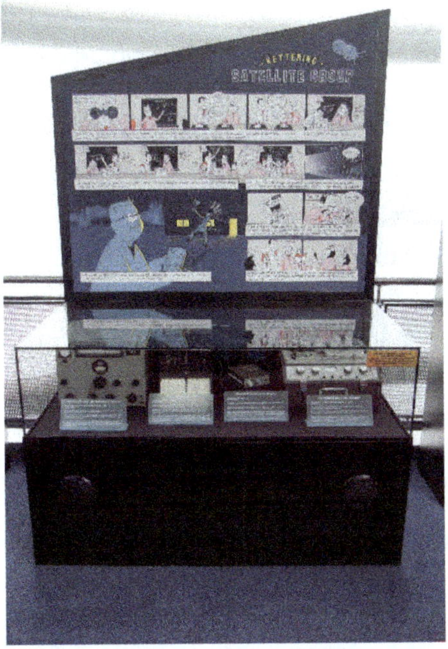

Kettering Group permanent exhibit at the National Space Centre.

Geoff's papers, equipment and documentation, and other materials had been preserved at his home in Cornwall after he died in 2000. Following the passing of Jean Perry 15 years later, Isabel assumed responsibility for Geoff's archive and relocated it to near her home in Herefordshire.

Doug Millard, Deputy Keeper of Technologies and Engineering at the Science Museum, tells the rest of the story:

> It was following a conversation with Bob Christy outside the Science Museum's 'Cosmonauts: Birth of the Space Age' (2015) exhibition that I was able to go on and meet Isabel (née Perry) and Charles Carmichael in 2016. Whilst Derek Slater had donated items, including his CR100 radio receiver, to the National Space Centre, Leicester, a substantial volume of varied material, including equipment, belonging to Geoff Perry was in the hands of his daughter, Isabel Carmichael. She was seeking an appropriate long-term custo-

dian for her late father's archive. To these ends, I confirmed with little hesitation that the Science Museum would eagerly provide such a home.

Two years later, and after a great deal of cataloguing help from Isabel, the final delivery of documents arrived at the Science Museum's Library and Archives. If I had to name one of my favourite archives, it would indeed be the 'Geoffrey Perry Archive relating to the operation of the Kettering Group.' It is wide-ranging, includes many different types of paper and record, and all created in Perry's meticulous style and detail. It is an immensely rich collection which offers a globally significant narrative of Soviet and Russian Federation satellite and probe launches. That fact that it also reflects a wonderful bridge between the lay and professional scientist only adds to its value; Perry operating from his school and involving his pupils while communicating, collaborating with and even informing his professional peers such as the Royal Aircraft Establishment's Desmond King-Hele.

The 'Geoffrey Perry archive relating to the operation of the Kettering Group,' (https://archives.sciencemuseumgroup.ac.uk/Details/archivescience/110093443) is available for inspection and research by appointment, which can be booked here: https://www.sciencemuseum.org.uk/researchers/dana-research-centre-and-library.[412]

Epilogue

Geoff wasn't ideological about his monitoring of the Soviet space programme. He said, 'I never considered myself a spy.' Then what was it about Soviet spacecraft behaviour that captivated Geoff and the Group? He said he enjoyed tracking Soviet satellites for two reasons: one, because of their higher orbital inclination, they were more visible in the UK than American satellites, and most Soviet spacecraft transmitted on a frequency that the Kettering Group's hardware could receive; and two, the Soviets divulged little about what their satellites' purposes were, hence in Geoff and the pupils they generated curiosity and a desire to crack the secrecy. The Americans were generally more open about their space missions' trajectories and purposes, so their satellites weren't as enticing. In short, said Geoff, 'We're interested in anything that presents a challenge.'[413] He also said of his activities, 'It's good fun, isn't it? The crossword puzzle syndrome, you could say. If the Soviets told us everything they did, I wouldn't be near as interested as I am.'[414]

Geoff was a leading example of 'citizen science'—that a group of people without formal qualifications in a particular specialty could make a contribution to a technical field. Not only were his pupils making observations and generating data that few, if any, of their contemporaries around the world were doing, but he also created a world-wide amateur network of contributors that fed him information to help shine a light on undisclosed aspects of international space programmes.

One of the first global scale citizen science projects was Operation Moonwatch, founded by the Smithsonian Astrophysical Observatory in 1956 in anticipation of the launching of artificial satellites. Volunteers who had no special technical skills were taught how to observe satellites in the night sky and how they record relevant data. Any information duly gathered was to be sent to the Smithsonian's network of sky-watching cameras to enable photography of any new spacecraft in orbit.

Today, a number of enthusiastic amateur satellite observers seek to understand the purpose of classified satellites and unravel their missions. The website satobs.org is a useful tool in helping them to achieve their objectives. The mailing list SeeSat-L on the satobs.org website allows observers to exchange information rapidly.

More broadly, citizen science efforts have multiplied rapidly in the last few decades. Most have internet hubs that anyone can peruse and decide how and when to contribute. Examples include Zooniverse, iNaturalist, and Scistarter. Many other more narrowly focused projects are available around the world, such as giraffespotter.org. The UK's Environment Agency works with local partners to facilitate public involvement in research design, data collection and interpretation. Governmental sites, like citizenscience.gov, provide people opportunities to work with organizations like NASA and NOAA.

What Geoff left behind was his teaching, his science and his pupils. Geoff's journey started with a simple desire—how can I better illustrate the Doppler effect to my pupils? Geoff and Derek made a recording of Sputnik IV demonstrating the effect, then started analysing Doppler shifts of early American satellites. The initial pupil trackers were young women, followed by more than 20 years of signal monitoring and analysis by boys from Kettering Grammar School and Kettering Boys' School. Many of Geoff's pupils went on to university and took science and engineering degrees.

The revelations and findings that Geoff and his pupil trackers divulged ultimately impacted the international geopolitical arena; neither the Soviets nor the CIA could hide the existence of a secret launch site near the Arctic Circle. The actions and findings of the Kettering Group showed how a small collection of curious people can make contributions to knowledge and affect the course of human affairs.

The Reverend Pauline Newton officiated at Geoff's memorial service, which was held at the Tresham Institute, Kettering, in September 2000. She is a friend of Isabel and Charles Carmichael and a bloodline relative of Sir Isaac Newton—his grandfather, Robert, is her ancestor, twelve generations back. Geoff's impact on revealing unseen truth and on science education in general was expressed succinctly by Reverend Newton in her 'welcome and introduction' portion of the ceremony:

> By its very nature this service will be looking back at past events, but it is important to keep at the forefront of our mind that the foundations Geoff laid have not disappeared. The knowledge, the skill, the discipline and the enthusiasm he generated will be carried on by others in a variety of ways and will be an invaluable legacy.[415]

Appendix A

Selected Kettering Group Members and Supporters

The pupil engagement within the Kettering Group existed from 1962 to 1984 across several year groups. Each year, about six pupils would be working with Geoff on satellite tracking projects. At the end of each academic year, a number of pupils in the Group left school to go on to higher education, so Geoff would invite younger pupils to join the Group for the following year. No official roster of the pupils' names exists; over the period of pupil engagement, some 40 to 50 individuals participated in the Group's activities.

The adult members of the Kettering Group, whether in the UK or overseas, maintained close contact with Geoff and data was shared, when appropriate, initially through correspondence and phone calls and later via email. As the international standing of Geoff and the Group grew, Geoff's interaction with many other people and organisations, often on an informal basis on specific topics, developed to become a substantial activity for him.

There is no definitive list of Kettering Group members; however, those who had regularly engaged with Geoff would be considered as members of the Group and in many cases received mementos of their association through their receipt from him, for example, of a tie or brooch bearing the Kettering Group logo.

Below are short profiles of those people closely associated with Geoff's satellite work. In several cases, some details are repeated from earlier chapters for completeness.

Jean Perry (née Brand)

Jean, Geoff's wife, was born in 1924 and grew up in Braintree, Essex, England.; On leaving school, she worked for her family's building firm. She and Geoff met through their shared love of music at their local church. Geoff proposed to her in the summer of 1953, and they married the following year, setting up home in Kettering. Jean managed and supported Geoff in his career. Isabel remembered, 'My mother worked tirelessly and was always busy in the home and tending the garden.'

Jean had a caring and retiring personality, always considering others' feelings. She was a good friend to Kettering Group members and their families and was always a good hostess whether in Kettering or in Bude. Isabel and Charles said, 'There is one member of the Group who deserves a special mention, namely Jean Perry. Jean was Geoff's rock, always by his side and supporting him in all that he did; although not in the spotlight, her contribution to his success should not be underestimated.'

Geoff said about Jean, 'My wife is very patient and long suffering, [and] she is a very good entertainer. She's a good organiser of telephone calls from reporters. She even processes the two-line [orbital] elements before I get them, putting ticks against the lines I'm interested in.'[416]

Following the sudden passing of Geoff in 2000, Jean lost both a husband and a way of life. She remained in Bude until 2013 when she was moved to be near Isabel and Charles in Herefordshire. Jean passed away in 2015.

Isabel Carmichael (née Perry)

Isabel was born to Jean and Geoff Perry on 17 March 1956, exactly ten years before the launch of Cosmos 112. She grew up in a household where satellite tracking was a normal and constant activity and indeed her whole life has been entwined with satellite monitoring and the work of the Kettering Group. From a young age. she answered the telephone and logged calls and messages; she had a standard response for a long-distance caller: 'My father isn't here at the moment. He can ring you back; will you accept if he reverses the charges?'[417] Geoff 'allowed' Isabel to record satellite passes in the logbook and Isabel remembered, 'During peaks of publicity, I would record the radio news bulletins using Daddy's reel-to-reel tape recorder.'[418] Isabel would assist Geoff at his lectures by operating the slide projector and tape recorder—such was their empathy, there was no need for 'next slide please'. She enjoyed the responsibility of her own project—monitoring the spin-rate of the UK's Prospero satellite in the early 1970s.

Constantly learning from her father, she developed a lifelong interest both in physics and teaching. On leaving Kettering High School, she read physics at the University of Reading, the same university her father had graduated from. After completing her Postgraduate Certificate in Education, her physics teaching career spanned 35 years. From 2006 she was Head of Science and Physics at Malvern St James Girls' School, formerly Malvern Girls' College. She keenly followed the orbital decay of Skylab in 1979 and the re-entry of the Salyut 7 space station in 1991, engaging her pupils in both projects.

Isabel married Charles in 1980 and they settled in Worcestershire. One of the memories of that special day recalled by Isabel: 'Being punctual was always of paramount importance, and just in case of traffic congestion or a lorry being stuck under the bridge in Northampton Road, Kettering, on my wedding day, Daddy and I had a 'pre-launch built-in hold' outside the Cattle Market en route to Rockingham Road Baptist Church, thus ensuring that we arrived 'spot-on.''[419] Isabel also wrote, 'I am immensely proud of my father and the scientific method I learnt from him underpins my life. I am grateful for all the experiences, the skills attained and friends I have made for life through the satellite work.'[420]

Isabel and Charles are retired and live in Herefordshire. Following her mother's passing in 2015, it was a privilege for her and Charles to organise her late father's extensive archive which is now held for the nation by London's Science Museum. Geoff was proud of Isabel, saying in 1986, 'Now she's a physics teacher like her father, teaching at Malvern Girls' College in England, one of the leading girls' independent schools. She got a better degree than I did from Reading University. She has a receiver and she's picking up weather satellites in the afternoon periods when the girls have free activities. So just as with the physics teaching where I've handed over my laboratory tricks to my daughter, the satellites are also carrying on the family business out in the west country.'[421]

Charles Carmichael

Charles and Isabel met at the University of Reading, where they were studying some of the same courses. Charles read Physics and Meteorology BSc followed by Applied and Modern Optics MSc. He worked for 30 years, first at the Royal Signals and Radar Establishment (RSRE), part of the Ministry of Defence Procurement Executive, and following its privatisation in 2001, QinetiQ plc until his retirement in 2008. Whilst at RSRE, Charles worked on thermal imaging and night vision technologies. In the latter part of his career, he was an account director responsible for developing business with leading international aerospace companies.

When Charles and Isabel stayed with Geoff and Jean, Charles recalled, 'Everything stopped for tracking passes. Something's coming by, everything stopped. Meals had to stop, or meals had to be at a time that didn't interfere with what was coming by next. If it was a launch or a docking for example, who would get the first signals? It was exciting waiting to learn of the mission's outcome.'[422] Charles always appreciated discussing the latest Kettering Group activities with Geoff and learning of any new and emerging Soviet capabilities from him—and, as Geoff would say, 'all good talking points.'

Isabel and Charles Carmichael.

Isabel and Charles are proud to be members of the Kettering Group, are grateful for the friendships made and simply cannot describe the immeasurable love, support and opportunities provided by Jean and Geoff throughout their lives.

Robert Christy

Robert Christy joined the Kettering Group just as the Plesetsk story was bringing Geoff and the Group worldwide attention. Christy left the Kettering Grammar School in 1969 and contributed to the Group's tracking activities for many years afterwards. He worked as an IT administrator and project manager for the local county council. He runs the website orbitalfocus.uk, specialising in space activities around the world. The website has an extensive section dedicated to the Kettering Group featuring a timeline of the activities of Geoff and the Group across several decades. Christy is retired and lives in the north of England.

On the Kettering Group, he said, 'We didn't realise we were learning. But when you look back 50 years later and you realise how many of the things—when I do something I can almost hear Geoff's voice telling me that's the way I should do it. When it comes to doing bits of analysis, I remember doing things with Geoff and doing it in certain ways.'[423]

Les Currington

A retired television and radio retailer, Les Currington specialised in receiving and printing images from Soviet weather satellites. He joined the Group after Geoff retired from teaching. Currington attended AMSAT UK conferences with Geoff and undoubtedly gave talks about his weather satellite monitoring. In 1990, Currington and Geoff examined space hardware at the Farnborough Airshow and asked about a new Chinese weather satellite. Having been met with shrugs from the Chinese representatives, they came back the next day and astonished them with images from the satellite that Les had received overnight.

Les was a keen gardener, and his wife Kay and Geoff's wife, Jean, were good friends. Les Currington passed away in 2008.

About listening to satellites, Currington said, 'If I hear an unusual sound, I'll come in and record it. Then if I can't identify it, I'll phone Geoff Perry, and he'll identify it.'[424]

David Dean

David Dean was a pupil member of the Kettering Group from 1971 to 1976. He obtained a PhD in nuclear physics from the University of Oxford. Dean remembered the Oxford admission interview process:

> Geoff said, 'Don't forget to take your satellite folder with you [to the interview].' I thought, this is an interview about studying physics, why should I take this thick folder with my notes? I decided to take it along and went to an interview with my future physics tutor....He said, 'I've looked at your application and see that you're from Kettering Grammar, have you done anything with the Satellite Tracking

> Group?' I said, 'Actually, I'm a member.' He said, 'Tell me', and I gave him my elevator speech, my three sentences that I had prepared as Geoff told me to. He was fascinated, he said, 'I'd really like to chat more.'...I went back, we spent an hour then going through stuff from my folder and I told him about the Group. I got in and he was my tutor for three years as an undergraduate, and he continued to support me during my PhD as well. So the work [with the Kettering Group] had a very specific role in helping me get into the college of my choice for my physics bachelors.[425]

Dean moved to Germany and worked as a senior partner at an international management consultancy. He is retired and lives in Germany. He remembered his time with the Group in the 1970s: 'Looking back, it was a special time in human history, and we had our own little role to play in it. Actually, I think the role was much bigger than we thought at the time....[Geoff] was an inspiration in so many ways. Without a doubt, he was one of the really special mentors in my life.'[426]

Stuart Eves

Stuart Eves worked on satellite systems, then joined the UK's Defence Intelligence Staff. Regarding his sensitive intelligence position, Eves said, 'One of the challenges I faced was that I was aware of material that was coming in that was classified. Obviously, Geoff had no clearance so I couldn't share that sort of information with him. He understood that. One thing that we did have fortunate access to was material that was being picked up at airshows...We were able to get brochures from airshows with pictures of Russian satellites and things like that. I would make a point of photocopying those and putting them in care packages that I would send down to Geoff.'[427] After many years at Surrey Satellites designing spacecraft, Eves now works as a space consultant and lives west of London. He was honoured to be an usher at Geoff's memorial service, along with Max White.

Prior to White's death, Eves said, 'Max White sends me emails on a regular basis. Quite often I'll get an email that says, 'Geoff would have loved this.' We do still hold him in very high esteem. I think he influenced both our careers quite significantly.'[428] Eves also recalled, 'Geoff once tipped me off that a [Space Shuttle] launch would be visible after sunset, and that we would be able to view the main fuel tank of the Shuttle as a separate object from the orbiter as they passed over the country. I duly made the observation and then dashed excitedly indoors to phone Geoff. "It was cloudy here" was the rather grumpy response I received from Bude.'[429]

Richard Flagg

When the first Sputnik satellite orbited the Earth in 1957, one of the ham radio hobbyists who received its signals was 17-year-old Richard Flagg. He was a radio astronomer at the University of Florida in the late 1960s when he wrote to Geoff about working with him. Geoff asked Sven Grahn to contact Flagg, and Grahn and Flagg are still good friends to this day. Flagg's chief interest was hardware; he built many receivers and other devices for Group members. His other main interest was listening to shortwave signals generated by the planet Jupiter. Flagg helped found the Radio Jove programme to encourage amateur radio operators and students to listen to signals from the gas giant; Radio Jove still exists and for many years used a textbook written by Flagg to educate its members.[430]

On Geoff and the Kettering Group, he said, 'Geoff is a particularly good teacher and relies on the basics, the very fundamentals of physics to teach young people. You need a focal point, and satellite tracking is an excellent focal point to understand a lot of the physics.'[431]

Richard Flagg, left, and Sven Grahn, Florida, 1978.

Sven Grahn

Sven Grahn is a shortwave radio enthusiast who was the first overseas member of the Kettering Group. He started listening to high frequency radio bands as a teenager. Grahn became fascinated by the orbiting Soviet satellites because, as he put it, 'US space activities were not secretive, while the Soviet space programme was enigmatic and cloaked in secrecy—irresistible for the curious!'[432] About his first satellite contact, and initial communication with Geoff, Grahn later recalled:

> I had read two articles in the British aviation magazine *Flight* [*International*] about a grammar school in the little town of Kettering in Northamptonshire that was listening in to satellites, so I wrote a letter to them and included a tape reel of my recorded signals. I got a prompt reply the following week—remember, this was in the days long before email, faxes and the like. In the letter Geoff Perry confirmed that I had indeed heard Cosmos 104, and that I should listen

> for another new satellite that had just been launched. He also said that I ought to get a stopwatch to make the reception times in my monitoring more accurate. This exchange of letters was the beginning of a 34-year friendship with Geoff. I had become a member of a loosely knit group of satellite listeners around the world.[433]

In a note to Desmond King-Hele a few months after Grahn first contacted him, Geoff wrote, 'This fellow in Stockholm is being pretty useful to us in many ways.'[434] Grahn became a key member of the Kettering Group and, partly because of his activities as a satellite tracker, studied engineering and worked in the Swedish space programme for several decades. He is retired and lives near Stockholm.

Grahn wrote of the Kettering Group, 'Our radios and computing methods were almost always at the 'trailing edge of technology', but I think Geoff regarded this as yet another pedagogical advantage. The use of simple methods puts emphasis on the deductive and 'basic physics' part of it all.'[435]

David Hawkins

David Hawkins met Geoff Perry at a meeting of satellite visual observers in the early 1970s. Geoff helped him install an HF antenna at his house and assisted him in making his first satellite radio observation. Hawkins specialised in 'two-tone' satellite signals, which encoded telemetry information using two separate frequencies. He was an industrial chemist by career and is retired and living in Hertfordshire.

He said, '[Geoff] is an enthusiast who gets excited about things. He is one of the people that I admire because like myself he is able to apply the science he was taught as a student to his everyday life and actually produce something worthwhile from it.'[436] He added, 'I certainly cherish the day I met Geoff. Like his pupils, I also went to a grammar school, and I wish he had been my physics teacher!'[437]

Jos Heyman

An accountant by trade living in Perth, Jos Heyman had a keen interest in spacecraft and rockets. For many years he edited the *News Bulletin of the Astronautical Society of Western Australia.* The Society invited Geoff to give a number of presentations in 1979; afterwards, Geoff contributed several articles to the *News Bulletin.* Heyman also ran his own spaceflight information company. Jos Heyman passed away in 2016.

Jos commented, '[In 1984] Geoff was interviewed for Soviet TV by Sergei Kapitsa (son of Pyotr Kapitsa, the Nobel Laureate in Physics) for his 'Science for Youth' programme. He said it was good fun telling the Russians frequencies on which they could listen to hear their own cosmonauts talking!'[438]

Desmond King-Hele

Desmond King-Hele was born the same year as Geoff Perry, 1927. King-Hele graduated with a degree in mathematics from the University of Cambridge in 1948. He was offered a position with the Royal Aircraft Establishment, Farnborough, and remained there for 40 years. He worked on guided missiles, satellite orbits, the structure of the atmosphere and geodesy—the size, shape and other characteristics of the Earth. He was a Fellow of the Royal Society, London, and he wrote several books about English poets such as Shelley, becoming a recognised scholar on the subject. He has a 21-part oral history available online at the British Library's website.[439] Desmond King-Hele passed away in 2019 at the age of 92.

In a 1966 letter to Geoff confirming the re-entry time of Cosmos 113 as being 0726 GMT, which Geoff had determined based on listening to the spacecraft's signal, King-Hele wrote, 'I can't remember whether I told you that we had a decay note from the [US Air Force] which gave a time agreeing with yours for Cosmos 113. They gave 0735 +/- 10 minutes, so you are the more accurate!'[440]

Len Maxim

Len Maxim worked at the Wellington Meteorology Office in New Zealand. He and Geoff communicated several times a year about observations Maxim made in the southern hemisphere. Maxim monitored meteorological satellites and made visual observations of Mir, Soyuz and Progress spacecraft. Len Maxim passed away in 2015.

Jonathan McDowell

As a child living outside of London, Jonathan McDowell was fascinated by all things space related. In the 1970s and 1980s, he attended BIS meetings and heard Geoff speak several times. McDowell earned a doctorate in astronomy from the University of Cambridge University and worked on several projects including the Chandra X-ray telescope at the Harvard-Smithsonian Center for Astrophysics. Geoff learned about his satellite-tracking database, *Jonathan's Space Report*, and contacted him through Marcia Smith. McDowell and Geoff exchanged faxes and emails for several years, discussing the details of various Russian missions. McDowell recently moved back to England after 37 years in the US.

In an interview, McDowell recalled, 'The biggest compliment I've ever received in my life is that one of the other Kettering folks told me that Geoff had said of me, 'Yes, he seems to be Kettering material.' Meaning of course he doesn't bullshit; he does the work and distinguishes what's known and what's not known.'[441]

Sarah Mobbs

Sarah Mobb's father owned a local butcher's business in Kettering where Jean Perry shopped. Mobbs remembered, 'Geoff would look in the window of the shop. I have a memory of talking to Geoff about a little radio I was [using] in the shop. Then Geoff said to my dad, "Do you want to bring Sarah along to the Grammar School?" [442] After working with Geoff on satellite tracking projects for several years, Mobbs trained to become a children's nurse specialising in neurological care. She remembered, 'One day when I said to Geoff, "I'm going to be a nurse," he looked at me and said, "What do you want to be one of those for, Sarah?" I said, "I've always wanted to do that." He said, "Do it!", as only Geoff could say it.' [443] Mobbs is retired and lives near Kettering.

Regarding Geoff's take on leadership, Mobbs recalled, 'I always remember Geoff saying to me, "Sarah," he said, "you don't need to know everything, you just need to know the people who can help you. That's what I try to do in the Kettering Group. I don't know everything, but I know somebody who can give me that information." That stuck with me.' [444] She also remembered how Geoff encouraged her. 'Geoff would say, "You're doing all right" and he would give me that confidence somehow. He used to bring that out of me. He obviously saw something.'[445]

Patrick Moore

Inspired by BIS Chairman and author Arthur C. Clarke, Patrick Moore started writing astronomy books in 1952 and edited the first two years of the BIS flagship publication, *Spaceflight*. He presented the BBC documentary television programme *The Sky At Night* for more than 50 years, and wrote over 200 books, mostly on astronomical topics.

Like Geoff, Moore was able to infer secret information from his observations. Whilst crossing the Atlantic during World War II, he deduced by observing stars in the night sky that the ship had steered north of its planned course, which was classified information as Moore found out from a very stern ship's captain; the ocean liner was avoiding German submarines, presumably using guidance provided by British intelligence.[446] Sir Patrick was knighted by Queen Elizabeth II in 2001. Patrick Moore passed away in 2012 at the age of 89.

Pierre Neirinck

Pierre Neirinck tracked satellites professionally at both RSRS and at the Royal Greenwich Observatory, and as a hobby before and after his professional tracking positions. Neirinck oversaw the Satellite Orbits Group at RSRS for many years. He provided Geoff and the Group with tracking data in the late 1960s and the 1970s. Neirinck was a colleague of Desmond King-Hele and wrote or contributed to several scientific papers on satellites and other celestial topics. He retired to France and continued his interest in satellite tracking until he passed away in 2016. After he retired, one of his colleagues arranged to have an asteroid named after him, 142014 Neirinck.

Ed O'Grady

Ed O'Grady worked at Eastern Airlines for 20 years and was a ham radio enthusiast. When he acquired his first personal computer in the 1980s, he found an application for tracking satellites and became a follower of the Soviet, then the Russian, space programme. O'Grady made several trips to Soviet aerospace design bureaus and to Baikonur, the Soviet Union's main launch complex. He finished his long career working at a car dealership in New York. O'Grady is retired and lives in the US on the coast of Georgia.

Isabel's pupils, who monitored the decay of Salyut 7's orbit in the early 1990s, wrote, 'The whole group would like to thank Mrs. Carmichael for all her help and encouragement. Also, many thanks to Geoffrey Perry and Ed O'Grady for supplying us with the latest Salyut 7 orbital data.'[447]

About talking with Geoff, O'Grady recalled, 'He would joke back and forth. They were just great discussions. I never spoke to Geoff where I didn't come away learning something. [He is] really, really, sorely missed.'[448]

Greg Roberts

Greg Roberts was a professor of astronomy in South Africa. He remembered 'the intense shock which resulted from the announcement by the Soviet Union that they had launched the world's first artificial satellite.'[449] Roberts was already fascinated by spacecraft and was monitoring them visually and by radio when he first contacted Geoff in 1978. From his southern location, Roberts was able to track certain non-Soviet satellites that couldn't be monitored in the UK. He corresponded regularly with Geoff for many years. Roberts is retired and lives near Cape Town.

He said, 'The [early] achievements of the Kettering Group were all the more remarkable in that they were done before the advent of microcomputers and in that everything was done on a shoestring.'[450]

Mark Severance

Mark Severance was invited into the Group by Geoff in early 1977, after starting satellite radio tracking with ASTP, China 2, Salyut 5 and numerous other satellites transmitting VHF telemetry. He was already regularly corresponding with Dick Flagg and Sven Grahn who provided the 14-year-old with copious technical advice and assistance with his radio tracking gear. The three have remained close friends over the years. He was fortunate to attend the only high school in his hometown of Fort Worth, Texas that offered Russian language classes, which he enrolled in because he wanted to understand what the cosmonauts were saying on the radio. He graduated from Southern Methodist University with degrees in electrical engineering and physics, and a minor in Russian.

Severance had a 30-year career as a NASA employee, spending his tenure in various technical and management positions in the Space Shuttle, NASA-Mir, and International Space Station (ISS) programmes. He supported real-time mission operations for the Space Shuttle and ISS at Mission Control Center-Houston and for NASA-Mir and early ISS operations as a NASA Operations Lead at Mission

Control Center-Moscow. At the time of his retirement in 2018, he was the Network Director for the NASA Human Space Flight Communications & Tracking Network. He and his wife Barbara, who also had a career at NASA as an engineer and astronaut trainer, live near the Sierra Blanca and Capitan mountains in south central New Mexico.

Severance said recently, 'I was very fortunate to become involved in the Group, which certainly guided me toward my educational and professional career paths. My Kettering Group roots grew many tendrils!'[451] He also said, 'Look at the number of lives that [Geoff] touched, and the amount of people that say Geoff is with them in their mannerisms and the way they approach things to this very day. My example is, don't mix your units! And don't speculate; stick with facts and data. If you can build a hypothesis from that, that's ok. But don't blindly speculate as to what you think something might be.'[452] And he added in a later remembrance, '[Geoff] was always teaching, even when you were just talking about stuff. You were always being taught by him, even if you didn't realise it. You may not realise it until later on in life. He was constantly in a teaching mode.'[453]

Charles Sheldon

Charles S. Sheldon II was Chief of the Science Policy Research Division of the Congressional Research Service, a unit of the U.S. Library of Congress that provides research and analysis exclusively for members and committees of Congress. A PhD economist, Sheldon joined CRS (then the Legislative Reference Service) in 1955 after serving in the U.S. Navy during World War II and the Korean War before transferring to the Naval Reserve and teaching at the University of Washington. After the launch of Sputnik in 1957, Sheldon was detailed to be the technical director for the House Committee on Science and Astronautics and helped write the 1958 National Aeronautics and Space Act that created NASA.

He later spent five years on the staff of the White House National Aeronautics and Space Council, participating in drafting the 1962 Communications Satellite Act that led to creation of the Communications Satellite Corporation (COMSAT) and the beginning of global satellite communications. Sheldon connected with Geoff following Geoff's public disclosure of the use and location of the Soviet Union's northern launch site, Plesetsk, in 1966 and visited Geoff while touring the UK in 1967. They were friends and colleagues for the rest of Sheldon's life.

Charles Sheldon, c. 1978.

Sheldon wrote extensively about the Soviet space programme for Congress, most notably the *Soviet Space Programs* 'green book' reports for 1966-1970 and 1971-1975. He was working on the 1976-1980 edition when he died in 1981 at the age of 64. Sheldon also assembled reports on other technical topics for the US House of Representatives and the US Senate. He wrote for publications including the *TRW Space Log* and *Air Force* magazine. For several years, he was a Distinguished Traveling Lecturer for the American Institute of Aeronautics and Astronautics, giving talks around the country about Soviet space activities.

Sheldon once said of Geoff, 'He's been able to gain an understanding of what the Russians are up to that very few other people, as individuals, perhaps no one else in the world, has.'[454]

Michael Sinnett

Michael Sinnett was a member of the Kettering Grammar School Satellite Tracking Group from 1963 to 1967. He lived next door to Geoff Perry in Kettering for many years. He remembered the day he moved into the house as a young boy and noticed Geoff. 'The first impression when I saw Mr. Perry through the window, I think it was, is that he was quite a frightening character, quite a big character. I thought, "Am I going to get on with him or not, really?"' [455] He added, 'It was Mr. Perry that took us into the back garden and got us interested in looking at stars and satellites and things going round the Earth, which to me, I've got no idea what they were or didn't exist to me before moving to Kettering, certainly.'[456]

Sinnett studied physics and spent most of his career in various information technology roles. He spent a few years instructing at a UK technical college and recalled, 'It was when I was teaching physics that I really went back to the Geoff Perry method. He used to come up with these wonderful examples to demonstrate things, stories that could [be used]. I used some of those tales when I was teaching the same class. That would have been in '78, so we're talking 15 years, probably, after Geoff had taught them to me. They were so strong they were still fresh in my mind.'[457] Sinnett is retired and lives near Sheffield.

He said of Geoff, 'This is a brilliant example of what a dedicated and enthusiastic teacher can achieve with his pupils. His lessons were always informative and fun. Those pupils lucky enough to be involved with the Satellite Tracking Group were taught the importance of accurate data recording and analysis to the success of any practical experiment. Most of the Group members, including myself, went on to study physics at university and we will always remember what we learnt from Mr. Perry.'[458]

Derek Slater

Derek and Geoff enjoyed a long and productive partnership and friendship. Derek later wrote, 'Geoff Perry and I got on famously from the start. We had two things in common: a conviction that the understanding of science is supremely important, in that it transforms and enhances our own enjoyment of the world, and secondly, that we were determined to build up the best school Science Department in the area, whatever it cost in effort. Apart from that, we had completely different personalities and temperaments. Still have. I tend to be casual, procrastinating, bored with fine detail and administration, excited by the next project before I've finished the first one. Geoff isn't.'[459]

Alongside his 26-year involvement with Geoff's satellite-tracking activities, Derek ran the school's Radio Club and, with Geoff, took pupils on tours of the European continent during holiday breaks. The friendship between Geoff and Derek, which began at school, extended to their respective families and endures to this day.

Derek Slater, c. 2010.

On his philosophy of teaching, Derek said, 'It has always been my belief that the processes of teaching and learning are interlinked. The day you stop learning, you may as well stop teaching.'[460] Derek retired from Kettering Grammar School in 1989, five years after Geoff. Later, he consulted and ran his own business for many years. Derek Slater passed away in March 2015. In Geoff's words, 'Without Derek's expertise, equipment and co-operation [the story of the Kettering Group] would never have come to be written.'[461] About how his involvement in satellite tracking started with Sputnik IV, Derek recalled (perhaps a bit tongue-in-cheek), 'If I'd known how it was going to affect my life for the next 25 years, I'd have taken more notice.'[462]

Marcia Smith

Marcia Smith joined the Science Policy Research Division of the Congressional Research Service as a space policy analyst in 1975. She wrote sections of the *Soviet Space Programs, 1971–75* report and took over coordination of the series within CRS after Charles Sheldon's death in 1981. She and Geoff collaborated in finishing the 1976-1980 edition and writing the 1981–1987 volumes. Geoff also contributed to Smith's annual *Space Activities of the United States and Soviet Union* reports throughout the 1980s and 1990s. She retired from CRS in 2006 and became Director of the Space Studies Board and of the Aeronautics and Space Engineering Board at the National Academies of Sciences, Engineering and Medicine, before starting her own consultancy in 2009. She is currently editor of spacepolicyonline.com, a website focused on space policy in the United States and abroad.

In 1984, Smith commented, 'The Kettering Group research has been extremely important and leading to a better understanding in the unclassified domain of what the Soviets have been doing. They are an excellent resource.'[463] She was honoured to be invited by Geoff to become a member of the Kettering Group and continues to this day to marvel at what he was able to accomplish with rudimentary equipment and a brilliant team of pupils and adults eager to learn from him and share his enthusiasm for solving mysteries.

Peter Wakelin

Peter Wakelin worked most of his career at the UK Meteorological Office. He was posted abroad many times, including assignments to the

Falkland Islands, Ascension Island, and Bahrain. Wakelin monitored Soviet meteorology and navigation satellites and occasionally Mir, Progress and Soyuz missions. He was also a prolific visual observer.

One of his fondest memories supporting the Kettering Group involved 'pedalling back home after night shifts at the Bahrain airport at great speed in summer heat dodging drivers and jumping red lights (everybody else does!) to hear TFs, TGs, and TKs [telemetry signals].'[464]

Max White

The orbital prediction service of the Radio and Space Research Station was transferred to Aston University in 1980. Max White was a member of staff at Aston University and helped generate data for the predictions. He provided tracking data to Geoff and the Group for the following 20 years. After operating tracking cameras in England and Australia for two decades, he changed careers and worked in finance but never lost his love for observing satellites, which started as a schoolboy. Max White passed away in 2024.

Max once said, of the data that Geoff readily provided to the organisations that helped him, 'Farnborough always appreciated all the work, all the little nuggets. [The RSRS at] Slough appreciated nuggets. Certainly, at Aston we did.'[465]

Chris Wood

Chris Wood, a distant relative of Geoff's, had been posted to Fiji by the UK General Post Office and was familiar with the operation of an HF radio. Geoff recruited Wood in the late 1960s as he was well located to receive transmissions Cosmos satellites; he was able to confirm that a particular Cosmos satellite was on its final orbit before re-entry. Wood and Geoff communicated via telegram using their customised encoded messages to relay tracking data and results. Upon returning to the UK, Wood developed an interest in Soviet navigation satellites and, with

Geoff's guidance, proved adept at decoding their signals. Chris Wood passed away in 2023.

He said, 'I believe that Geoff Perry is a born leader. This is reflected in the way he's organising the Group and particularly in the way he is respected by his pupils at school and the way he has been able to lead them in very important activities.'[466]

Appendix B

Pulse Duration Modulation Telemetry: ON/OFF Keying Versus Frequency Shift Keying & Doppler Tracking Techniques for Both

By Mark Severance

The original Soviet telemetry system used a simple ON/OFF keying concept, similar to how amateur radio operators transmit Morse code. The value of each bit of transmitted data, was represented by the time duration or the 'pulse' of the radio signal, i.e., the time the transmitter was keyed ON. At the end of a pulse the transmitter was briefly turned off until the start of the transmission of the next bit. The time duration of the ON pulse represented the value of the data bit being telemetered. Fifteen telemetry data bits were transmitted in this manner, constituting a single telemetry frame, during a time duration of just over 16 seconds.

The end of a telemetry frame, contiguous with the beginning of the next frame, was signaled by a transmission of a rapid train of very short ON/OFF pulses, the pulse train lasting just over 1.5 seconds. Further analysis by Geoff Perry and Richard Flagg determined that the elapsed time durations between the start of the first pulse in the pulse train and the time between subsequent pulses in the train was used to define the possible transmitter ON times of the sequent data bit, thus quantizing the possible values of the data bits in the telemetry frame. Hence the name 'Pulse Duration Modulation telemetry (PDM)'. An idealized waveform a single frame of PDM of telemetry is shown in Fig 1B.

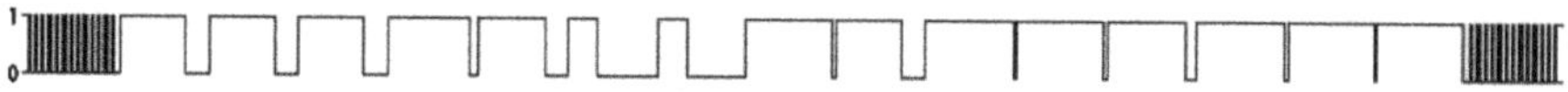

Figure 1B – Idealized PDM telemetry wave form.

A new system was later introduced that used a two-frequency method. Instead of a data bit value corresponding to the time of a pulse transmission on a single frequency, the system shifted pulse transmissions between two frequencies approximately 1 kHz apart. The time duration of the transmitter pulse on the higher frequency (fH) represented the actual value of the data bit being telemetered. At the end of that bit the transmitter switched to the lower frequency (fL) until the start of the transmission of the next data bit. At the start of the next bit, the transmitter shifted back to fH for the time duration representing the value of that data bit.

Similar to the ON/OFF keying scheme, 15 telemetry bits were transmitted in this manner, constituting a single telemetry frame, over a time duration of approximately 16 seconds. Also similar to the original system, at end of a telemetry frame/beginning of the next frame, a rapid pulse train was generated by shifting the transmitter very quickly between fH and fL , with the pulse train lasting approximately 1.5 seconds. Analogous to the PDM ON/OFF keyed system, the elapsed time durations between the frequency shifts from the first pulse of the train on fH and the subsequent pulses on fL quantized the possible values of the subsequent data bits in the telemetry frame. Hence the name 'Frequency Shift Keying-Pulse Duration Modulation telemetry (FSK-PDM).'

After gaining an initial understanding of the two-frequency FSK-PDM system, the question of how best to conduct Doppler tracking to determine the time of closest approach was addressed by Geoff and Derek.

For the single frequency ON/OFF keyed system, Doppler tracking was straight forward. The Doppler effect (also known as Doppler shift) is apparent due to the motion of the satellite relative to the ground station. A graphic showing a typical curve of this Doppler shift is shown in Figure 2B.

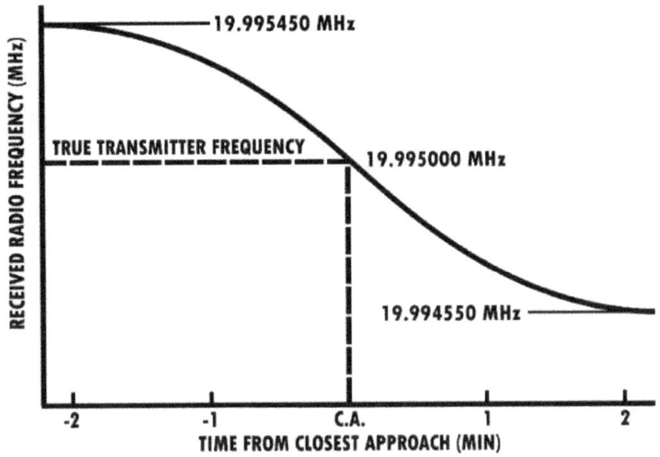

Figure 2B – Typical Doppler shift curve of a transmitted signal from a satellite as it passes by a ground station.

When the satellite first appears above the ground station horizon and is approaching the station, the received satellite signal appears higher than the actual transmitted frequency and the received signal frequency decreases as the satellite nears its time of closest approach to the ground station. The point of closest approach is reached at the time when the rate of frequency change is at its maximum. Also at this time, the true frequency (that is the frequency actually transmitted by the satellite) is received. As the satellite travels away from the ground station, the received frequency continues to decrease to values below the transmitted frequency.

The introduction of the two-frequency FSK system required the development of a new tracking technique to measure the Doppler shift. Geoff and Derek noted that the 1 kHz separation of fH and fL frequencies was approximately equal to the Doppler shift observed at 20 MHz. They realized that Doppler shift curves for both frequencies could be obtained by setting the receiver's beat frequency local oscillator (BFO) to the true (non-Doppler-shifted) frequency of fH. A graphic representation of this frequency configuration and the resulting Doppler shift is provided in Fig 3B.

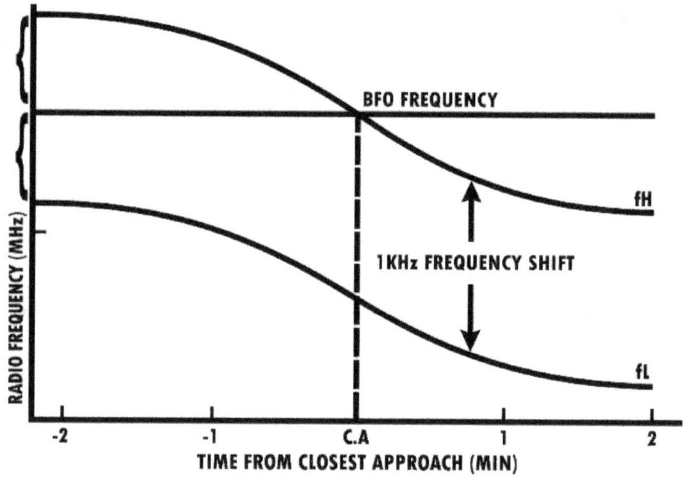

Figure 3B – Two- frequency Doppler shift curves.

This setting was made by adjusting the BFO early in the pass, while the satellite was approaching, until two the audio tones of the "beat frequency" (the resulting audio tones from the mixing of the BFO signal with the fH and fL signals) sounded roughly equivalent. The presence of equivalent beat frequency audio tones indicated that the BFO frequency was tuned to the fH.

As the pass progresses the audio frequency of the beat frequency tones change. The changes in the audio tones are the result of the mixing of between the two Doppler shifted fH and fL frequencies and the fixed BFO frequency. The audio frequency response of the beat frequency audio tones due to Doppler shift is shown in Fig. 4B.

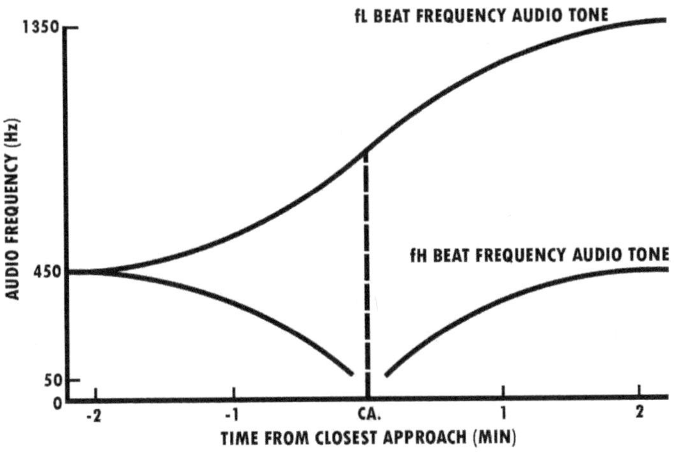

Figure 4B – Audio frequency response of beat frequency due to Doppler shift.

The difference between fH and the fixed BFO frequency decreases as the satellite approaches, therefore the beat frequency audio tone also decreases. Identical to the Doppler shift seen from the single frequency system, the rate of frequency change of fH and fL and the rate of change of the beat frequency audio tones increases as the satellite nears its point of closest approach. When the beat frequency audio tone drops below 30 Hz, the beat frequency audio tone corresponding to fH becomes inaudible and passes into the 'zero-beat condition,' i.e., the fH signal received from the satellite is within 30 Hz of the fixed BFO frequency (and the actual frequency transmitted by the satellite), indicating the satellite has effectively reached the point of closest approach.

The zero-beat condition typically remains for approximately 40 seconds for satellites in the standard Cosmos-Zenit orbit on high elevation passes above the ground station. The received fH frequency continues to decrease as the satellite passes beyond the point of closest approach. When the frequency difference becomes greater than 30 Hz, a beat frequency audio tone re-appears out of the zero-beat condition and continues to increase in frequency for the remainder of the pass. The mid-time of zero-beat condition can be estimated as the actual the time of closest approach, providing an accuracy of less than +/- 30 seconds.

Since fL is below the fixed BFO oscillator frequency and the received fL decreases throughout the pass due to Doppler shift, the beat frequency audio tone generated from fL increases in frequency throughout the pass.

By using the observations from these Doppler shift tracking techniques, the orbital ground track of the satellite could be calculated. With these data, the launch and landing times of Cosmos/Zenit satellites could also be determined using the algorithm described in Appendix C.

Appendix C

An Algorithm for Launch Time, Ground Track and Landing Time Calculations for Cosmos/Zenit Satellites

By Mark Severance

Geoff's algorithm was based on several observed facts and data:

- The inclinations of all satellites launched by the Soviets were available through TASS announcements of the launch or in the RSRS bulletins.
- The 8-day Cosmos/Zenit reconnaissance satellites generally flew in orbits with 65-degree inclinations.
- An 8-day Cosmos/Zenit reconnaissance satellite would nominally deorbit for a landing on revolution 128.
- At the nominal altitude of 8-day Cosmos/Zenit reconnaissance satellite, the satellite would complete 16 orbits in a 24-hour period.
- Observed times of the closest approaches of the satellite determined from the Doppler tracking of its HF telemetry signal.

With this information, Geoff's calculation algorithm was as follows (Steps 1 to 7 applied to a launch from Baikonur, while step 8 shows modifications needed for a Plesetsk launch):

1. The elapsed time between close approaches on consecutive days, as determined from Doppler tracking, is divided by 16 (the number of orbits completed by a Cosmos/Zenit satellite in a 24-hour period). This gives the orbital period within +/- 0.1 min. Knowing the orbital period, the satellite's orbital altitude and track separation could be calculated; track separation being the difference in longitude between consecutive northbound equator crossings (ascending nodes) of the satellite's ground track.

2. An ascending node satellite orbit reaches its maximum latitude at a value equal to the orbital inclination of the satellite, the so-called "north point" of the orbit. The north point is a specific point on the satellite's ground track, therefore for a specific ascending node orbit there is a specific longitude corresponding to the north point.

3. For a satellite in a 65-degree inclination orbit on a ground track which passes nearly overhead at Kettering (i.e. a pass with maximum elevation of approximately 90 degrees), the point at which the 65-degree north point latitude is reached corresponds to a longitude of 52 degrees East.

4. Calculating back from the observed time of closest approach (adding eight minutes to account for the time from closet approach to reach the north point) and using the north point longitude of 52 degrees East, the orbital period and track separation are repeatedly subtracted until the launch date is reached.

5. For 65-degree inclination launches from Baikonur, the north point of the initial orbit occurs at longitude 125 degrees East. When the back calculations of ascending node longitudes on the launch date reach the range of 113.5-136.5

degrees East, this orbit and ascending node time are considered to be the initial "Rev 0", i.e. the hypothetical full-orbit ground track corresponding to the launch of the satellite.

6. The longitude is then set to 125 degrees East. Calculating the time from the Rev 0 ascending node to point along the ground track where the satellite reaches 45 degrees north latitude (the latitude of the Baikonur Cosmodrome), the satellite's launch time can be estimated with reasonable accuracy.

7. To estimate the landing time, forward calculations are made to by repeatedly adding the orbital period and track separation to the calculated Rev 0 ascending node time and longitude (125 degrees East). By making these calculations until Rev 128, the typical deorbit rev for 8-day Zenit missions, the landing time can be estimated.

8. After estimating an approximate location of the Plesetsk Cosmodrome, Geoff's calculations showed that for a satellite launched from Plesetsk in a 65 degree inclination orbit would reach its Rev 0 north point at a longitude of 60 degrees East. Similarly, the Rev 0 ascending node time and Rev 128 landing time can be estimated by using this value instead of 125 degrees East. The back calculated Rev 0 ascending node time would correspond to an ascending node longitude range of 48.5-71.5 degrees. Calculating the time from the Rev 0 ascending node to the point along the ground track where the satellite reaches 63 degrees north latitude (the latitude of the Plesetsk Cosmodrome), the satellite's launch time can be estimated with reasonable accuracy.

ENDNOTES

The abbreviation KET refers to the Kettering Group archives at the Science Museum's Science and Innovation Park.

Introduction – The Soviets and the CIA

[1] Asif Siddiqi, *Sputnik and the Soviet Space Challenge* (University Press of Florida, 2003), 129.

[2] Ibid., 160-161.

[3] Matt Bille and Erika Lishock, *The First Space Race* (Texas A&M University Press, 2004), 63.

[4] William E. Burrows, *This New Ocean* (Modern Library, 1998), 168-171, 177-181.

[5] Siddiqi, *The Soviet Space Challenge,* 155; Bille and Lishock, *First Space Race*, 66; Philip Clark, *The Soviet Manned Space Programme* (Orion Books, 1988), 9-10.

[6] Bart Hendrickx, 'Building a Rocket Base in the Taiga: The Early Years of the Plesetsk Launch Site (1955-1969)–Part 1,' *Space Chronicle*, vol. 65, Supp. 2 (2012): 52.

[7] Amy Ryan, and Gary Keeley, 'Sputnik and US Intelligence: The Warning Record,' *Studies In Intelligence*, vol. 61, no. 3 (2017): 6.

[8] Ibid.

[9] Martin Caidin, *Man Into Space* (Pyramid Books, 1961), 168.

[10] Evert Clark, 'Soviet Technical, Political Gains Spur Shift in Attitude on Defense,' *Aviation Week and Space Technology*, 21 October 1957: 27.

[11] W. H. Lawrence, 'President Voices Concern On US Missiles Programme, But Not On the Satellite,' *The New York Times*, 10 October 1957: A1.

[12] *Soviet Capabilities in Guided Missiles and Space Vehicles* (CIA, 1959), 2-5.

[13] Philip Klass, 'How the US Taps Soviet Missile Secrets,' *Aviation Week and Space Technology*, 21 October 1957: 26-27.

[14] *Possible Guided Missile Test Ranges in the USSR* (CIA, 1957), 15.

[15] Rolf Tamnes, *The United States and the Cold War in the High North* (Dartmouth Publishing Company, Ltd, 1991), 135.

[16] Associated Press, 'Space Capsule Is Caught in Mid-air By US Plane on Re-entry from Orbit,' *The New York Times*, 20 August 1960: A1.

[17] Dwayne Day (editor), *Eye in the Sky: The Story of the CORONA Spy Satellites* (Smithsonian Books, 1998), Introduction, 'Corona.'

[18] Chris Pocock, Dragon *Lady: The History of the U-2 Spyplane* (Motorbooks International, 1989), 37.

[19] Walter F. Hahn, *Internal Motives for Soviet Secrecy* (Institute for Defense Analyses, 1963), 2.

[20] Ibid., 6.

[21] Ibid., 30, 40-41, 47-50.

[22] Jeffrey Richelson and William Arkin, 'Spy Satellites: 'Secret,' But Much Is Known,' *Washington Post*, 6 January 1985: C-1.

[23] Ibid.

Chapter 1 – Secret Base, Secret Satellite

[24] 'Cosmos 112, Probable Photorecce Craft, First Space Launch From Plesetsk,' *NORAD Weekly Intelligence Report*, 25 March 1966: 10; https://www.archives.gov/files/declassification/iscap/pdf/2009-068-doc119.pdf

[25] Desmond King-Hele et al, *The RAE Table of Earth Satellites, 1957-1986* (Stockton Press, 1987), 103.

[26] James E. David, 'Starting to Crack a Hard Target,' National Security Archive, 5 February 2020, https://nsarchive.gwu.edu/briefing-book/intelligence-nuclear-vault/2020-02-05/starting-crack-hard-target-us-intelligence-efforts-against-soviet-missile-programme-through-1957

[27] Olav Riste, *The Norwegian Intelligence Service, 1945-1970* (Taylor and Francis, 2014), 149.

[28] Stanley G. Zabetakis & John F. Peterson, 'The Diyarbakir Radar,' CIA, 1996, https://www.cia.gov/resources/csi/static/The-Diyarbkir-Radar.pdf.

[29] Author's simulation of Cosmos 112 orbit using Systems Took Kit v. 12.1 with a two-line element derived from prediction bulletins for Cosmos 112 from the Radio and Space Research Station (RSRS) and the Space Detection and Tracking System (SPADATS) Center.

[30] David Dean, interview, April 2024.

[31] Author's simulation of Cosmos 112 orbit.

[32] G. E. Perry, 'The Kettering Group,' *Space Communication and Broadcasting*, vol. 5 (1987): 405-407; G. E. Perry, 'A school satellite tracking station as an aid to teaching physics,' *Physics Education*, vol. 3 (1968): 286; Isabel Carmichael, 'Satellite Tracking in Kettering and beyond—my recollections,' Kettering Civic Society talk, 14 October 2016.

[33] Geoff Perry, 'The Extra-Curricular Satellite Studies,' in *Cytringanian Farewell.* (Old Cytringanians, 1995), 243.
[34] Author's simulation of Cosmos 112's orbit.
[35] Ian Poole, 'Radio Waves and the Ionosphere,' *QST,* November 1999: 62.

Chapter 2 – Geoff, a Portrait

[36] Perry, *Cytringanian Farewell,* 235.
[37] Isabel Carmichael, interview, June 2023.
[38] Robert Christy, interview, August 2023.
[39] Geoff Perry, interview with Jos Heyman, part 1, 1986, http://www.sven-grahn.pp.se/trackind/Ketterin/Interviews_with_Geoff_Perry_1986.htm
[40] Perry, *Cytringanian Farewell,* 235
[41] Arthur C. Clarke, 'Extraterrestrial Relays,' *Wireless World,* October 1945: 305-308.
[42] Perry, *Cytringanian Farewell,* 237.
[43] Isabel and Charles Carmichael, interview, June 2023.
[44] Ibid.
[45] Ibid.
[46] 'ITN Special: Apollo-Soyuz Special,' Getty Videos, 1975, https:// www.gettyimages.fi/detail/video/apollo-soyuz-special-england-studio-intlive-presenter-uutismateriaali/1289023484?adppopup=true
[47] Mark Severance, interview, February 2024.
[48] Sarah Mobbs, interview, September 2023.
[49] Geoff Perry, letter to Desmond King-Hele, 27 February 1966, KET/J/38.
[50] Letter to Geoff Perry, courtesy of Isabel Carmichael.
[51] Sarah Mobbs, interview, September 2023.
[52] Isabel and Charles Carmichael, interview, April 2024.

Chapter 3 – A New Sound

[53] Perry, *Cytringanian Farewell,* 235.
[54] Slater, *Cytringanian Farewell,* 243.
[55] G. E. Perry, 'Listening to Satellites,' *Proceedings of the Royal Institution,* vol. 60 (1990): 15.
[56] John Gau, 'Sputniks, Bleeps and Mr. Perry, Part 2,' Channel 4, television programme, 1987. Reposted on YouTube by MikeyKett, video, 10:17, https://www.youtube.com/watch?v=hDmF7bYCZyo.

[57] Patrick Moore, 'The Kettering Connection,' BBC Radio 4, broadcast May 12, 1985. Reposted by Martin Mobberley on youtube.com, audio, 28:05, https://www.youtube.com/watch?v=1mk6mSoSjBo.

[58] Perry, *Cytringanian Farewell,* 237.

[59] Derek Slater, 'Members' Tales,' Orbital Focus, accessed 10/3/2022, https://www.orbitalfocus.uk/Kettering/DerekS.php.

[60] 'Local men 'followed' spaceship,' *Kettering Evening Telegraph*, 16 May 1960: 6.

[61] Files for 1961-027A Discoverer 32, 1961 Alpha Gamma 1 and 1961-034A Discoverer 36, 1961 Alpha Kappa 1, KET/C/3/03.

[62] Perry, *Cytringanian Farewell,* 237.

[63] Geoff Perry and Derek Slater, 'A Quantitative Doppler effect Experiment for the Grammar School Sixth Form.,' *Spaceflight*, November 1962: 198-200; G. E. Perry, 'Tracking Russian Satellites,' *Wireless World*, vol. 73, no. 3 (1967): 132.

[64] 'Signals from Outer Space: Analysis of telemetry from Sputnik 1 and alternate ways of obtaining information,' *NSA Technical Journal*, vol. III, no. 2 (1958): 77-82, https://nsarchive2.gwu.edu/NSAEBB/NSAEBB501/docs/EBB-01.pdf

[65] Geoff Perry and Derek Slater, 'Quantitative Doppler effect Experiment,' Appendix 6, in *A Handbook of Astronautics.* (Dufour Editions, 1963), 89-90.

Chapter 4 – The First Pupil Satellite Trackers

[66] Larrie D. Ferreiro, *Measure of the Earth: The Enlightenment Expedition That Reshaped Our World* (Basic Books, 2011), Introduction.

[67] Desmond King-Hele, letter to Geoff Perry, KET/C/9/29.

[68] Geoff Perry, 'To determine the Earth's flattening from the rotation of the orbital plane of an Earth satellite,' *School Science Review*, vol 43, no. 149 (1961): 176-177.

[69] Robert Christy, interview, August 2023.

[70] Desmond King-Hele, 'The effect of the Earth's oblateness on the orbit of a near satellite,' *Proceedings of the Royal Society A: Mathematical, Physical and Engineering Sciences*, vol. 247, (1958): 49.

[71] Geoff Perry, 'The determination of a satellite's orbital period from many simultaneous observations,' *School Science Review*, vol. 46 (1964): 138-142.

[72] Perry, *Cytringanian Farewell,* 238.

[73] Perry, *Cytringanian Farewell,* 239.

[74] Ibid.; Barbara Jacques, G. E. Perry, and J. D. Slater, 'Cosmos Observations,' *Flight International,* 16 July 1964: 97.

[75] King-Hele et al, *Table of Earth Satellites*, 52.

[76] 'Kettering Children Listen In As Two Astronauts Chat,' *Kettering Evening Telegraph*, 10 July 1964: 5

[77] Ibid.

[78] Perry, interview with Jos Heyman, part 1.

[79] Kenneth Owen, 'Kettering's Cosmos Scholars,' *Flight International*, 22 July 1965: 145.

[80] G. E. Perry and J. D. Slater, 'Cosmos Observations,' *Flight International*, 12 November 1964: 842.

[81] R. A. Jahn, 'The Outer Radiation Belt,' Flight International, 6 May 1965: 714.

[82] Richard Jahn, 'Members' Tales,' Orbital Focus, accessed 21 September 2021, https://www.orbitalfocus.uk/Kettering/RichardJ.php

[83] Perry, 'A school satellite tracking station,' 283.

[84] Howard Miles (ed.), *Artificial Satellite Observing* (Faber and Faber Limited, 1974), 133-135.

[85] Geoff Perry, 'Tracking Satellites At School', *New University*, January 1969: 12.

[86] Ibid.

[87] Gau, 'Sputniks, Bleeps and Mr. Perry,' Part 2.

[88] Slater, *Cytringanian Farewell*, 245.

Chapter 5 – Cosmos 112 and Plesetsk

[89] Peter A. Gorin, 'Zenit: Corona's Soviet Counterpart,' in *Corona: Between the Sun and the Earth*, Robert McDonald (editor), (The American Society for Photogrammetry and Remote Sensing,1997), 103.

[90] Philip Klass, *Secret Sentries in Space* (Random House, 1971), xiii.

[91] Alan Thomson, 'US Naval Space Command Space Surveillance System,' Federation of Atomic Scientists, accessed 24 November 2023, https://spp.fas.org/military/programme/track/spasur_at.htm.

[92] Zabetakis and Peterson, *Studies in Intelligence*, 41-3.

[93] Jeffrey Richelson, *The Wizards of Langley* (Westview Books, 2002), 79.

[94] 'Morning meeting of 18 March 1966,' CIA, 19 December 2016, www.cia.gov/readingroom/document/cia-rdp80b01676r001500090050-0

[95] Ibid.; 'KH-7, GAMBIT-1', Gunter's Space Page, https://space.skyrocket.de/doc_sdat/kh-7.htm#:~:text=gambit%201

[96] Author's simulation of the Cosmos 112 orbit.

[97] P. Joyce, 'A Bow Tie, A Thermos Flask and Wheelbarrows,' in *Cytringanian Farewell*, Old Cytringanians (Old Cytringanians, 1995): 234.

[98] Isabel and Charles Carmichael, interview, June 2023.

[99] Geoff Perry, 'The Soviet Northern Cosmodrome,' *Spaceflight*, August 1967: 274.

[100] Geoff Perry, 'Observing Earth Satellites review,' *Spaceflight*, vol 9, no. 3 (1973): 107.

[101] Ibid.

[102] John Gau, 'Sputniks, Bleeps and Mr. Perry, Part 3,' Channel 4, television programme, 1987; reposted on YouTube by MikeyKett, video, 13:10, https://www.youtube.com/watch?v=YlG1XYZqjuk&t=666s

[103] Perry, interview with Jos Heyman, part 1.

[104] Perry, *Soviet Northern Launch Site*, 245.

[105] Ibid.

[106] Bart Hendrickx, 'Building a rocket base in the taiga: the early years of the Plesetsk launch site (1955-1969)–Part 1,' *Space Chronicle*, Supplement 2 (2012): 58.

[107] Ibid.

[108] 'CORONA 9013 mission report,' CIA, 21 December 1961, 8. https://www.cia.gov/readingroom/document/cia-rdp78t05693a000300010011-7

[109] Tamnes, *Cold War in the High North*, 136.

[110] *Soviet Technical Capabilities in Guided Missiles and Space Vehicles* (CIA, 1961), 7, https://www.cia.gov/readingroom/docs/DOC_0000278407.pdf

[111] Hendrickx, 'Building a rocket base in the taiga: the early years of the Plesetsk launch site (1955-1969)–Part 2,' *Space Chronicle*, Supplement 1, 2013, p. 16.

[112] Hendrickx, 'Building a Launch Site–Part 2,' 15.

[113] Gorin, *Corona Between the Sun and the Earth*, 92.

[114] Ibid.

[115] Ibid., p. 93-94.

[116] *CIA's Analysis of the Soviet Union, 1947-1991*, CIA (1992), 136-141.

[117] *The Corona Story*. National Reconnaissance Office, 1988, Appendix B-9, https://apps.dtic.mil/sti/tr/pdf/ADA640829.pdf.

[118] Hendrickx, 'Building A Launch Site–Part 1,' 56, 68.

[119] Hendrickx, 'Building A Launch Site–Part 2,' 2-4, 16.

[120] Gorin, *Corona Between the Sun and the Earth*, 93

[121] 'Cosmos 113 Also a Photorecce Satellite' and 'Plesetsk May Specialize in Military Space Launches,' *NORAD Weekly Intelligence Report*, 25 March 1966: 10-13.

[122] Perry, 'The Soviet Northern,' 275.

[123] Geoff Perry, 'New Cosmos Launch Site?' in 'Geoffrey E. Perry MBE and the Kettering Grammar School Satellite Tracking Unit, ca. 1960-1984', Doug Millard, paper at 2019 International Astronautical Congress, Washington, D.C.

[124] Associated Press, 'Soviets Launch Satellite, Mum on Rendezvous Talk', *Tampa Bay Times*, 21 March 1966: 6.

[125] Perry, 'The Soviet Northern', 275.

[126] Ibid., 246.

[127] Geoff Perry, 'New Soviet Launch Site Pinpointed,' *Flight International*, 10 November 1966: 817.

[128] Gau, 'Sputniks, Bleeps and Mr. Perry,' part 3.

[129] Joseph Myler, 'British Schoolboys Spot Soviet Spaceport.' *Washington Post*, 13 December 1966: A1.

[130] Alfred Friendly, 'Satellite-Tracking British School Lab 'Beats' Space Signals Into Tone Effects,' *Washington Post*, 19 December 1966: A1, A8.

[131] Philip Klass, 'Cosmos Betrays New Launch Site,' *Aviation Week and Space Technology*, 19 December 1966: 16-17.

[132] Herbert Coleman, 'Kettering Boys Use War Surplus Gear,' *Aviation Week and Space Technology*, 19 December 1966: 17.

[133] 'Mars Voyage Plan Ends in Budget Pit,' *The Times*, 20 December 1966: 6.

[134] 'Schoolboys Track Satellites,' British Pathe, 1966, https://www.youtube.com/watch?v=QWY4nm1bmUU

[135] Gau, 'Sputniks, Bleeps and Mr. Perry,' part 3.

[136] 'Pentagon Signals a Hit!' *Kettering Evening Telegraph*, 21 December 1966: 1.

[137] Evert Clark, 'A 3D Missile Base in Use in Russia.' *The New York Times*, 21 December 1966: 25; 'British Pupils Track Satellites Using WWII Equipment,' *The New York Times*, 21 December 1966: 25.

[138] 'Kapustin Yar,' *Pravda*, 6 June 1983, translation by Mark Severance.

[139] 'Place of Launch—Plesetsk,' *Pravda*, 20 June 1983, translation by Mark Severance.

[140] 'Ask Any Schoolboy.' *New Scientist*, 29 December 1966: 716.

[141] Alan G. Mencher, letter from US Embassy, 19 January 1967. KET/J/10.

[142] Sven Grahn and Bjorn Gimli, 'New Fix on Soviet Launch Site.' *Flight International*, 26 January 1967: 150.

[143] Max White, interview, December 2023.

Chapter 6 – Evolving Capability and Membership

[144] 'Great Northern Undulator Model 310 Instruction booklet,' https://www.commsmuseum.co.uk/publications/GNT/GNT310.pdf ; Great Northern Telegraph Undulator, paper tapes, KET/E.

[145] Perry, *Cytringanian Farewell,* 239.

[146] Robert Christy, 'The Schoolboy's Tale of Satellite Tracking,' *Sven's Space Place*, accessed 22 May 2022, www.svengrahn.pp.se/trackind/Ketterin/Bob.htm, from *Space Flight News*, May 1987.

[147] 'Gift To Kettering', *Aviation Week and Space Technology*, 24 April 1972: 24.

[148] 'Ariadne,' *New Scientist*, 18 January 1969: 147.

[149] 'Kettering Report,' *Aviation Week and Space Technology*, 22 April 1968: 18.

[150] David Dean, interview, April 2024.

[151] 'They Spot Satellites,' *The Young Observer*, 19 May 1972.

[152] Geoff Perry, 'Radio Observations at Kettering Grammar School,' *Spaceflight,* vol. 14, no. 12 (1972): 478.

[153] Sven Grahn, 'Cosmos Vision' *Spaceflight*, vol. 17 (1975): 440-441.

[154] 'Spaceflight at 60,' *Spaceflight*, October 2016: 382.

[155] R. D. Christy, 'Safety Practices for Soyuz Recoveries,' *Spaceflight*, vol. 23 (1981): 321-322.

[156] Desmond King-Hele, *A Tapestry of Orbits* (Cambridge University Press, 1992), 138.

[157] Perry, *Cytringanian Farewell*, 240.

[158] Letter from the Soviet Embassy to Geoff Perry, 6 April 1977, KET/C/9/30.

[159] Geoff Perry, 'Listening To Satellites,' *Rotary Magazine UK*, August 1977.

[160] G. E. Perry and Sarah M. Mobbs. '74 deg, 95 min Cosmos satellites.' *Spaceflight* 22 (1980): 38-39.

[161] Sarah Mobbs, interview, September 2023.

[162] Ibid.

[163] Geoffrey Perry, 'Pupil Projects Involving Satellites,' *Space Education*, vol. 1, no. 7, (1984): 320-323.

[164] 'Industry Observer,' *Aviation Week and Space Technology*, 30 July 1984: 13.

[165] 'The Kettering Connection,' 1985.

[166] Stuart Eves, interview, October 2023.

[167] Perry, *Cytringanian Farewell*, 235-242.

[168] Ibid., 238.

Chapter 7 – Early Telemetry Analysis

[169] Perry, *Cytringanian Farewell*, 237; G. E. Perry, J. D. Slater and K. J. Edwards, 'Periodic Fluctuations in the Signals from Cosmos 5,' *Spaceflight*, 1967, vol. 9, no. 2: 66-68.

[170] G. E. Perry, 'The Effect of Solar Control of the Ionosphere on Reception of Signals from 65° Cosmos Satellites During Recovery,' *Spaceflight*, 8.8 (1966): 284.

[171] G. E. Perry, J. D. Slater, J. Marshall, S. Grahn, 'Cosmos Satellite Switching Sequences,' *Spaceflight, vol* 8, no.12 (1966): 435.

[172] G. E. Perry, 'Tracking Russian Satellites,' *Wireless World*, vol 73, no. 3 (1967): 132.

[173] G. E. Perry and S. Grahn, 'Inflight Frequency Changing of Cosmos Satellites,' *Spaceflight*, December 1968, 431-433.

[174] G. E. Perry, 'Two Recent Cosmos Recoveries,' *Flight International*, 95, 22 May 1969: 844-845.

[175] Stuart Eves, interview, October 2023.

Chapter 8 – Photo-Reconnaissance Spacecraft

[176] Sven Grahn, *Dawn, This is Golden Eagle* (Instant Book, 2019), 28-29.

[177] Ibid.

[178] Ibid.

[179] G. E. Perry, and S. Grahn. 'Recovery Beacon Signals from Cosmos Satellites.' *Spaceflight, vol.* 104 (1968): 142.

[180] G. E. Perry, 'Reconnaissance aspects of 8-day Cosmos satellites?', *Spaceflight* 10 (1968): 204-206.

[181] Saunders Kramer, 'Soviet Satellites,' *Aviation Week and Space Technology*, 16 February 1970: 43.

[182] G. E. Perry, 'Soviet Satellites,' *Aviation Week and Space Technology*, 6 April 1970: 102.

[183] G. E. Perry, 'Recoverable Cosmos Satellites,' *Spaceflight*, Vol 14: 84.

[184] 'Zenit Satellites – Zenit 4M variant,' Orbital Focus, accessed 14 June 2023, https://www.orbitalfocus.uk/Diaries/Zenit/Zenit-4M.php

[185] G. E. Perry, 'Cosmos Coverage of the Indo-Pakistani War,' *Spaceflight*, September 1972: 350.

[186] Stuart Eves, interview, October 2023.

[187] G. E. Perry, 'Recoverable Cosmos Satellites with Scientific Missions,' *Spaceflight*, Feb 74: 69.

[188] G. E. Perry, 'Looking Down on the Middle East War,' *Flight International*, February 21, 1974: 240, 245.

[189] Craig Covault, 'New Russian Recon keyed to area surveillance,' *Aviation Week and Space Technology*, 23 February 1998: 37.

[190] The Kettering Group, 'Russia launches an opto-electronic reconnaissance satellite into unique orbit,' *The Imaging Science Journal*, vol 46 (1998): 45.

[191] G. Perry, 'And whoever heard about the projects Zerkalo, Nord and Arkon-1?' *News Bulletin of the Astronautical Society of Western Australia* (1998): 27-29.

Chapter 9 – Human Spaceflight

[192] G. E. Perry, 'Zerkalo,' *Journal of the British Interplanetary Society*, vol. 52 (1999): 348-349.

[193] 'Boys Hear Cosmos 186,' *Aviation Week and Space Technology*, 6 November 1967: 17.

[194] Peter Fairley, 'Schoolboys Tune in to a Secret: Russia Sends Up Giant New Spaceship,' *Evening Standard*, 28 October 1967: 1.

[195] 'Radio observations of Kosmos186/188— the first automatic docking,' Sven's Space Place, http://www.svengrahn.pp.se/trackind/K186188/K186188.html

[196] Peter Fairley, 'Russia Plans Space Station 1,' *Evening Standard*, 3 November 1967: 1.

[197] R. S. Flagg and G. E. Perry, 'Telemetry from Russian Spacecraft,' *Journal of the British Interplanetary Society*, vol. 23 (1970): 451-464.

[198] Geoff Perry, 'USSR Space Station Research?' *Flight International*, 10 December 1970: 923. There is no author attribution for the article, but Charles Sheldon attributes it to Perry *in Soviet Space Programs, 1971-75,* on p. 187.

[199] Grahn, *Dawn, This Is*, 45.

[200] 'Three Soviet cosmonauts on way to docking with space station,' *The Times*, 7 June 1971: 1.

[201] *The Times Educational Supplement*, 11 June 1971.

[202] Clark, *Soviet Manned Soviet Space*, 64; Grahn, *Dawn, This is*, 47.

[203] 'Russia Orders Inquiry Into why three cosmonauts died,' *The Times,* 1 July 1971: 1, 6.

[204] Moore, 'The Kettering Connection.'

[205] 'School Detects Launch of Two Spacemen,' *The Times*, 4 July 1974: 1.

[206] Geoffrey Perry, 'Another Kremlin Gremlin?' *The Guardian*, 29 August 1974: 11.

[207] Clark, *Soviet Manned Space*, 70.

[208] Grahn, *Dawn, This is,* 67-69.

[209] James R. Asker, 'Soviets Plan Daily for US Emergency Landings,' *Aviation Week and Space Technology*, 22 July 1991: 22.

[210] Sven Grahn, 'Future Salyut Missions,' *Spaceflight*, October 1974: 392-393.

[211] The Kettering Group, 'Reception of radio signals from Cosmos 557,' *Spaceflight*, vol 16, (1974): 39-40.

[212] Max White, interview, December 2023.

[213] Richard Flagg, interview, November 2023.

[214] A.J. Driver and G.E. Perry, 'Salyut 6 Manoeuvres,' *Spaceflight*, vol. 21, 1979: 371.

[215] Clark, *Soviet Manned Space Programme*, 99-100.

[216] 'Docking Canceled for Soviet Craft,' *The New York Times*, 22 April 1983: A4.

[217] 'Soyuz Failure Laid to Error in Rendezvous,' *Aviation Week and Space Technology*, 2 May 1983: 21.

[218] Clark, *Soviet Manned Space Programme*, 134.

[219] 'Cosmonauts Complete Transfer Between Mir and Salyut 7,' *Aviation Week and Space Technology*, 12 May 1986: 28.

[220] 'Cosmonauts in Second Day Aboard Mir Space Station,' *Aerospace Daily*, 10 February 1987: 211-212.

[221] 'Soviets Dock Module to Mir after Aborted Attempt,' *Aviation Week and Space Technology*, 13 April 1987: 27.

[222] 'EVAs, Station Changes Keep Cosmonauts Busy,' *Aviation Week and Space Technology*, 24 July 1995: 63.

Chapter 10 – Individual Mission Analyses

[223] Philip J. Klass, 'Cosmos Flights Seen ICBM Aids,' *Aviation Week and Space Technology*, 16 October 1967: 26-27.

[224] G. E. Perry, 'Cosmos Flights,' *Aviation Week and Space Technology*, 6 November 1967: 102.

[225] G. E. Perry, 'Telemetry from the First Chinese Artificial Earth Satellite,' *Nature*, vol. 226 (1970): 829.

[226] Perry, 'Listening to Satellites,' 28.

[227] G. E. Perry, 'Observations of a second Chinese satellite,' *Space Communication and Broadcasting*, vol. 5, (1987): 397-403.

[228] Flagg, R. S. 'Telemetry from Chinese satellites.' *Space Communication and Broadcasting* 5 (1987): 385-395.

[229] 'China's new return capsule,' *Aviation Week & Space Technology*, vol. 137, no. 7 (1992): 13.

[230] 'PRC Failure Creates Risk,' *Aviation Week and Space Technology*, 25 October 1993: 71.

[231] 'Re-entry predictions for Cosmos 398, FSW-1 5 and TSS-1R,' ResearchGate, accessed 18 April 2024, https://www.researchgate.net/publication/234205603_Re-entry_predictions_for_Cosmos_398_FSW-1_5_and_TSS-1R

[232] Perry, interview with Heyman, part 1.

[233] G. E. Perry, J. D. Slater, I. J. Perry, 'Measurements of spin rate of the Prospero satellite,' *Journal of the British Interplanetary Society*, vol. 26, no. 3 (1973): 167-169.

[234] King-Hele, *Tapestry of Orbits*, 119, 121, 153; David Baker, *The Rocket* (Crown, 1978), 171.

[235] Geoff Perry, 'Venus Probe Failure,' *Flight International*, 13 April 1972: 513.

[236] 'Cosmos 482 Identified As Failed Venus Mission,' *Aviation Week and Space Technology*, 10 April 1972: 15.

[237] 'Failed Soviet Venus lander Kosmos 482 crashes to Earth after 53 years in orbit,' Space.com, 10 May 2024, https://www.space.com/space-exploration/launches-spacecraft/failed-soviet-venus-lander-kosmos-482-crashes-to-earth-after-53-years-in-orbit

[238] New Zealand Department of Scientific and Industrial Research, Report on Ashburton balls, 1972, KET/C/9/82.

[239] Geoff Perry, letter to Geoff Page, 16 September 1972. KET/C/9/82.

[240] Andy Salmon, 'Express Delivery', *Spaceflight*, vol. 39, January 1997: 7,8.

[241] Geoff Perry, 'What Really Happened to Express,' *News Bulletin of Astronautical Society of Western Australia*, vol. 21, no. 2 (1995): 17-19.

[242] Letter to Geoff Perry from Claus Lippert, 21 March 1996, KET/C/4/16/13.

[243] Stuart Eves, personal communication, February 2026.

Chapter 11 – Satellite Networks

[244] Geoff Perry, 'Cosmos at 74 deg,' *Flight International*, 30 November 72: 789-91.

[245] Geoff Perry, 'Identification of a navigation satellite system within the Cosmos programme,' *Journal of the British Interplanetary Society*, vol. 29, no. 5 (1977): 307-316.

[246] Kenneth Gatland, 'A Thousand Cosmos Satellites,' *New Scientist*, 13 April 1978: 92.

[247] Moore, 'Kettering Connection.'

[248] C. D. Wood and G. E. Perry, 'The Russian satellite navigation system,' *Philosophical Transactions of the Royal Society of London*, vol. 294, no. 1410 (1980): 307-315.

[249] P. Daly and G. E. Perry, 'Recent developments with the Soviet Union's VHF satellite navigation system,' *Space Communication and Broadcasting*, vol. 4 (1986): 51-61.

[250] P. Daly and G.E. Perry, 'Update on the behaviour of the Soviet Union's VHF satellite navigation system,' *Space communication and broadcasting*, vol. 5, no. 5 (1987): 379-384.

[251] G. E. Perry, 'Operational trends in Russian navigation satellite systems,' *Space Policy*, vol. 13, no. 1 (1997): 5-7.

[252] Michael R. Gordon, 'Mir Shadow Falls on Russian Role in Space,' *The New York Times*, 2 August 1997: A1, A5.

[253] Philip Klass, *Secret Sentries in Space* (Random House, 1971), 218.

[254] Richard D. Lyons, 'Soviet Satellite Destroyer Is Believed to Be in Orbit,' *The New York Times*, 6 February 1970: A1.

[255] 'Comments on Soviet Satellite Destroyer Article in the The New York Times, February 6 1970,' *Journal of the Office of Legislative Counsel*, 9 February 1970. CIA Reading Room, https://www.cia.gov/readingroom/document/cia-rdp72-00337r000100040034-7

[256] Mark Severance, interview, February 2024.

[257] R. J. Atwood and G. E. Perry, 'Cosmos 699 Explodes,' *Flight International*, May 1, 1975: 718.

[258] Charles Sheldon, editor, *Soviet Space Programmes 1971-75*, (US Government Printing Office, 1976), 432. Sheldon confirms here that the article titled 'Cosmos 699 Explodes' was written by Atwood and Perry.

[259] Ibid.

[260] G. E. Perry, 'Statsionar Engineering Test,' *Flight International*, April 4, 1974: 439.

[261] G. E. Perry, 'The Molniya Communications Satellites,' *Royal Air Forces Quarterly*, Summer, 1977: 59.

[262] Perry, *Cytringanian Farewell*, 240.

[263] G. E. Perry, 'Second Molniya 3 Launched,' *Flight International*, 24 April 1975: 686.

[264] Charles Sheldon, 'The Soviet Space Programme,' *Air Force Magazine*, vol 58, no 3 (1975): 50-56.

[265] Anatoly Zak, 'US-K and US-KMO constellations,' last updated 14 November 2024, www.russianspaceweb.com/oko.html ; Stuart Eves, personal communication, March 2026.

[266] G. E. Perry, 'Detection of the malfunction of Cosmos 954 and Cosmos 1402 and the subsequent prediction of their re-entries,' 1. KET/C/4/10/02.

[267] Geoff Perry, 'Track Recorders,' *The Listener*, 5 April 1990: 5.

[268] Gus W. Weiss, 'Life and Death of Cosmos 954,' *Studies in Intelligence*, Spring 1978: 2.

[269] Perry, 'Detection of the malfunction,' 1.

[270] Frank Miles, 'If It's Geoff On The Phone, It's A Hot One!' *The Lens*, March 1983: 3.

[271] Bernard Gwertzman, 'Nuclear Powered Soviet Satellite is Expected to Crash This Month', *The New York Times*, 6 January 1983: A1.

[272] Ted Koppel, 'Nightline', ABC, video, 5 January 1983.

[273] Alan Copps, 'Nuclear Satellite Breaks Up,' *Daily Telegraph,* 8 January 1983: 1.

[274] Geoffrey E. Perry, 'Perestroika and glasnost in the Soviet space programme: A personal view,' *Space Policy*, November 1989: 282.

[275] Geoffrey Perry, 'Soviet ELINT Satellites Cover the Globe,' *Military Electronics/Countermeasures*, vol. 9 (1983): 38.

[276] Geoffrey Perry, 'War Watchdogs Spying From the Skies,' *Space Technology International 1992*: 27.

[277] Ibid., 29.

[278] G. E. Perry, 'Factors affecting the choice of orbits for observation satellites,' *Journal of Photographic Science*, vol. 42, no. 2 (1994): 63

Chapter 12 – Wide-Ranging Activities

[279] Donald Forbes, 'Schoolboy Space Snoopers Spy on Satellites,' *Waterloo [Iowa] Daily Courier*, 30 June 1971: 32.

[280] G. E. Perry, 'Progress In Space,' *Physics Bulletin*, vol. 22, no. 3, (1971): 167.

[281] Richard D. Lyons, 'Experts Say Russia Plans Manned Landing on Moon,' *The New York Times*, 6 September 1971: A33.

[282] King-Hele, *Tapestry of Orbits*, 124.

[283] Max White, interview, September 2023.

[284] Ibid.

[285] Ibid.

[286] 'Space Research: Observing Satellites,' *Nature*, vol. 246 (1973): 32.

[287] King-Hele, *A Tapestry*, 218.

[288] G. E. Perry, 'No Rookie,' *New Scientist*, 20/27 December 1984: 79.

[289] Geoff Perry and Cliff Ranft, 'Soviet Satellite Longevity,' *Aviation Week and Space Technology*, 20 October 1986: 160.

[290] Desmond Ball, 'Soviet Signals Intelligence (SIGINT),' Research School of Pacific Studies, Australian National University (1989): 122-123.

[291] H. W. Miles and G. E. Perry, 'Visual observation of a water dump from the Space Shuttle,' *Journal of the British Astronomical Association*, vol.102, no.1 (1992): 7.

Chapter 13 – Television, Radio and Print

[292] Kenneth Owen, 'Kettering's Cosmos Scholars,' *Flight International*, 22 July 1965: 144.

[293] Ibid., 145.

[294] Perry, *Cytringanian Farewell*, 237.

[295] 'School Satellite Tracking Station.' Video, ATV, 1966, https://www.macearchive.org/films/atv-today-18031966-school-satellite-tracking-station

[296] Michael Sinnett, 'Members' Tales.' Orbital Focus, January 2016, https://www.orbitalfocus.uk/Kettering/MikeS.php

[297] 'Panorama,' Video, BBC One, broadcast 16 December 1968.

[298] Christopher Ward, 'Scooped by the 4th Form,' *Daily Mirror*, 9 June 1971: 10.

[299] Geoff Perry, 'ITN Special: Apollo-Soyuz Special,' Getty Videos, 1975, https://www.gettyimages.fi/detail/video/apollo-soyuz-special-england-studio-intlive-presenter-uutismateriaali/1289023484?adppopup=true

[300] 'Historic space rendezvous over Bognor,' *Bognor Regis Observer*, 18 July 1975.

[301] 'Television review,' *Financial Times*, 23 July 1975.

[302] Frank Miles, letter to Geoff Perry, 25 July 1975. KET/C/6/15.

[303] Duncan Campbell, 'Spy Skywatcher,' *New Scientist*, 11 May 1978: 364.

[304] Ibid., p. 365.

[305] Ibid., p. 365.

[306] Ibid., p. 366.

[307] R. W. Apple, Jr, 'British Teacher Tunes In On Space Shots,' *The New York Times*, 27 April 1980: 8E.

[308] Ibid.

[309] Ibid..

[310] 'Listening In', Video, *60 Minutes*, 21 February 1982, CBS.

[311] Stephen Solomon, 'Amateurs Eavesdrop on Soviet Satellites,' *Science Digest*, January 1984: 32.

[312] Ibid., p. 36.

[313] Ibid., p. 67.

[314] Peter Fairley, 'An Open Letter of Farewell to Geoff Perry', 1 December 1997, KET/L/02.

[315] Moore, 'The Kettering Connection.'

[316] Ibid.

[317] Ibid.

[318] Ibid.

[319] Ibid.

[320] Ibid.

[321] 'Amateur Sleuths Track Soviet Space Program,' *The New York Times*, 18 March 1986: 56.

[322] Ibid.

[323] Ibid.

[324] Nancy Banks-Smith, 'The Wizard of Kettering,' *Arts Guardian*, 25 March 1987: 12.

[325] Ibid.

[326] Gau, 'Sputniks, Bleeps.'

[327] Tony Smith, 'How the Boys Scooped the World,' *Northampton Evening Telegraph*, c. 1995. KET/J/02.

[328] Michael Sinnett, interview, November 2023.

[329] Geoff Perry, letter to Phil Redmond, 8 April 1990. KET/C/4/10/20.

[330] Doug Stewart, 'Satellite Sleuths,' *Air and Space/Smithsonian*, June/July 1990, p. 28.

[331] Ibid., p. 34.

[332] 'Russian Space Mirror,' video, Getty Images, 1993, https://www.gettyimages.com/detail/video/russian-space-mirror-englandintcms-geoffrey-perry-the-news-footage/868880354

[333] Robert Christy, 'The Kettering Group—Space Explorers with Their Feet Firmly on the Ground,' *Space Chronicle*, vol. 67, Supplement 1 (2014): 13.

Chapter 14 – Travels

[334] Geoff Perry, 1976 trip diary, KET/C/9/80.

[335] Jos Heyman, 'Members' Tales,' Orbital Focus, c. 2020, https://www.orbitalfocus.uk/Kettering/JosH1.php

[336] Severin Crisp, 'Visit of Mr. Geoffrey Perry, MBE,' *The Australian Physicist*, vol. 24, no. 2 (1987): 51.

[337] Marcia Smith, 'Remembrance and Thanks,' Memorial Service for Geoffrey Edward Perry MBE, 11. KET/L/22.

[338] Max White, interview, December 2023.

[339] Budpendra Jasani, editor, *Outer Space: A New Dimension of the Arms Race*, (Taylor and Francis, 1982), v.

[340] Joseph Rotblat and Sven Hellman, *Nuclear Strategy and World Security*, (McMillan Press, 1985), 377.

[341] 'Pugwash Conferences on Science and World Affairs,' The Nobel Prize, accessed 5/18/2023, https://www.nobelprize.org/prizes/peace/1995/pugwash/facts/

[342] 'Late Bloom For Crystal Gardens,' *New Scientist*, 2 January 1993, http://www.newscientist.com/article/mg13718541-900-late-bloom-for-crystal-garden/

[343] Geoff Perry, 'I Saw Aussat Go!' *News Bulletin of the Astronautical Society of Western Australia*, September/October 1985: 12.

[344] Martin Davidoff, *Satellite Experimenter's Handbook*, (American Radio Relay League, 1990), 14-15.

[345] 'Colloquium,' *Satellite News*, AMSAT-UK, August 1995.

[346] Isabel and Charles Carmichael, interview, June 2023.

[347] Max White, interview, December 2023.

[348] Rodney Buckland, 'Space Reaches Into the Classroom,' *Space Policy*, February 1991: 81.

[349] Letters to Geoff Perry, KET/L/01.

Chapter 15 – Author, Contributor, Editor

[350] Charles Sheldon (editor), *Review of the Soviet Space Program with Comparative United States Data*, (US Government Printing Office, 1967): 4.

[351] James Noble Wilford, 'Charles S. Sheldon II Obituary,' *The New York Times*, 12 September 1981: 16.

[352] Sheldon, *Review of the Soviet Space Program*, 5.

[353] Donald C. Winston, 'House Unit Reports Soviet Payload Gains,' *Aviation Week and Space Technology*, 4 December 1967: 98.

[354] Charles Sheldon (editor), *Soviet Space Programs, 1966-1970*, (US Government Printing Office, 1971): 176.

[355] Kenneth Gatland, 'Artificial Satellite Observing,' *New Scientist*, 1974.

[356] Isabel and Charles Carmichael, interview, June 2023.

[357] Charles Sheldon, editor, *Soviet Space Programs, 1971-1975* (US Government Printing Office, 1976): 440.

[358] Ibid., 477-478.

[359] Barbara A. Luxenberg (editor), *World-Wide Space Activities* (US Government Printing Office, 1977), vii.

[360] Charles Sheldon and Marcia Smith, editors, *Soviet Space Programs: 1976-1980, Vol. 1* (US Government Printing Office, 1982): v.

[361] Max White, interview, December 2023.

[362] Nicholas Johnson, *The Soviet Year In Space,* 1985 (Teledyne Brown Engineering, 1986), iii.

[363] Nicholas L. Johnson, *Europe and Asia in Space 1993-1994* (Kaman Sciences Corporation, 1995), 424.

[364] Davidoff, *Experimenter's Handbook*, 14-13.

[365] Geoff Perry, *The Kettering Table of Earth Satellites*, KET/K/1/2.

[366] Perry, 'Perestroika and glasnost,' 279.

[367] Ibid., 286, 287.

[368] Dominic Phelan, editor, *Cold War Space Sleuths*, (Springer Praxis, 2013), 7.

[369] Arthur Smith, *Planetary Exploration* (Patrick Stevens, 1988), 7.

[370] King-Hele, *Tapestry of Orbits*, 73.

[371] Marcia Smith, *China's Space Program—A Brief Overview Including Commercial Launches of US-Built Satellites*, (Congressional Research Service, 1998), 1.

Chapter 16 – Influence on Education

[372] Geoff Perry, 'Satellites and Astronomy,' *Physics Education*, vol. 1, no. 3 (1966): 207.

[373] Kevin Keohane, 'Physics Education—21 years on,' *Physics Education*, vol. 22, no. 3 (1987): 143.

[374] John Percy, 'The Kettering Project,' *Crucible: The Journal of the Science Teachers Association of Ontario*, vol. 4, no. 7 (1973): 26, 27, 29.

[375] J. C. Codling, 'School Projects in Space Physics,' *Spaceflight*, February 1974: 70.

[376] Ian Ridpath, 'Putting Students Into Space,' *New Scientist*, 16 February 1978: 410.

[377] H. Paul Schuch, 'Demonstrating Celestial Mechanics Through Measured Doppler Shift,' *QEX*, April 1992: 5.

[378] Thomas Becker, 'Mission to Earth and global space education policy,' *Space Policy*, May 1992: 154.

Chapter 17 – Honours and Awards

[379] Sven Grahn, speech, Geoff Perry memorial service, 30 September 2000, Tresham Institute, Kettering.

[380] 'Presidential Addresses on Awards,' *Quarterly Journal of the Royal Astronomical Society*, vol 15 (1974): 222.

[381] 'Annual Awards Citations, 14 December 1984,' Royal Aero Club, 2-3. KET/L/09.

[382] Moore, 'Kettering Connection.'

[383] 'Teachers of Physics,' *Physics Bulletin*, vol. 37, no. 5 (1986): 194.

[384] Geoff Perry, speech, Open University Honorary Degree, 6 May 1989, National Space Centre Collections, https://www.spacecentre.co.uk/collections/categories/kettering-satellite-tracking-group/kettering-group-memo-gp8901/

[385] 'Geoff gives name to college science centre,' *Northampton Evening Telegraph*, 3 September 1999: 19.

[386] 'Technology at the Tech—For the 21st Century,' *Herald Tribune* (Torquay, Devon, England), 11 June 1990: 16.

[387] 'Laurels 99,' *Aviation Week and Space Technology*, 7 February 2000: 25.

[388] 'Geoffperry,' Minor Planet Physical Properties Catalogue 4 December 2023, https://mp3c.oca.eu/bodies/Geoffperry/;

'Kettering,' Minor Planet Physical Properties Catalogue, 4 December 2023, https://mp3c.oca.eu/bodies/Kettering/

Chapter 18 – Legacy

[389] 'Mr Geoffrey Perry: Obituary,' *The Times*, 21 January 2000: 27.

[390] David Whitehouse, 'Space Tracking Pioneer Dies,' BBC News, http://news.bbc.co.uk/2/hi/science/nature/611875.stm

[391] James Glanz, 'Geoffrey E. Perry Dies at 72; Monitored Soviet Satellites,' *The New York Times*, 22 January 2000: A13.

[392] James Oberg, 'British schoolteacher, cracked secrecy of Soviet space programme, dies,' UPI, https://www.upi.com/Archives/2000/01/20/British-schoolteacher-cracked-secrecy-of-Soviet-space-programme-dies/6553948344400/

[393] 'Obituary,' *Aviation Week & Space Technology*, vol. 152, no. 4 (2000): 21.

[394] 'Geoffrey Edward Perry,' *Spaceflight*, vol. 42, no. 4 (2000): 165.

[395] Ibid.

[396] Dan Brekke, 'Quiet Passing of an Unlikely Hero,' *Wired*, 22 January 2000, https://www.wired.com/2000/01/quiet-passing-of-unlikely-hero/

[397] 'Geoff Perry, SK,' *Jonathan's Space Report*, no. 418 (2000), planet4589.org/space/jsr/back/news.418.txt

[398] 'A Place In History,' *Space News*, 31 January 2000: 14.

[399] Arthur C. Clarke, *The Exploration of Space* (Pocket Books, 1979): 233.

[400] David Dean, speech, Geoff Perry memorial service, 30 September 2000, KET/L/22.

[401] Derek Slater, speech, Geoff Perry memorial service, 30 September 2000, KET/L/22.

[402] Sven Grahn, speech, Geoff Perry memorial service, 30 September 2000, KET/L/22.

[403] Marcia Smith, speech, Geoff Perry memorial service, 30 September 2000, KET/L/22.

[404] Jonathan McDowell, interview, October 2023.

[405] Sven Grahn, interview, August 2023.

[406] Rex Hall and David J. Shayler, *The Rocket Men* (Springer Praxis, 2001): xix.

[407] 'Geoff Perry,' Project Kettering, 3 June 2008, archived 25 May 2024 at https://web.archive.org/web/20240525180344/http://www.projectkettering.org.uk/page_id__79_path__0p3p20p.aspx

[408] Ibid.

[409] http://www.svengrahn.pp.se; https://www.orbitalfocus.uk

[410] Derek Slater, 'Science Fantasy,' *The Times*, 1 September 2011, https://www.thetimes.com/uk/science/article/science-fantasy-gdtzkr5jdgh

[411] Tim Peake, 'Kettering Grammar School message,' Facebook, 16 March 2016, https://www.facebook.com/ESATimPeake/posts/1114521908587594

[412] Doug Millard, personal communication via Isabel and Charles Carmichael.

Epilogue

[413] Stewart, 'Satellite Sleuths,' 34.

[414] Apple, Jr., 'British teacher', 8E.

[415] Rev. Pauline Newton, speech, 2000 Geoff Perry memorial, Kettering, KET/L/22.

[416] Perry, interview with Jos Heyman, part 1.

[417] Isabel and Charles Carmichael, interview, June 2023.

[418] Ibid.

[419] Isabel Carmichael, speech, 2000 Geoff Perry memorial, 30 September 2000, Kettering. KET/L/22.

[420] Isabel Carmichael. Speech: Perry, Physics, Plesetsk, Prospero and Philately. Personal communication with the author.

[421] Geoff Perry, interview with Jos Heyman, 1986.

[422] Isabel and Charles Carmichael, interview, June 2023.

[423] Robert Christy, interview, August 2023.

[424] Stewart, 'Satellite Sleuths,' 30.

[425] David Dean, interview, April 2024.

[426] Ibid.

[427] Stuart Eves, interview, October 2023.

[428] Ibid.

[429] Stuart Eves, personal communication.

[430] NASA's Radio Jove project: https://radiojove.gsfc.nasa.gov

[431] Richard Flagg, interview, October 2023.

[432] Grahn, *Dawn, This is*, 9.

[433] Grahn, *Cytringanian Farewell*, 250.

[434] Geoff Perry, letter to Desmond King-Hele, 18 May 1966, KET/L/01.

[435] Grahn, *Dawn, This is*, 17.

[436] Dave Hawkins, interview, KET/H/5.

[437] Dave Hawkins, personal communication, November 2023.

[438] Jos Heyman, 'Geoff Perry Watches Space For ITN,' *News Bulletin of the Astronautical Society of Western Australia*, November 1984: 6.

[439] https;//docslib.org/doc/350275/desmond-king-hele-interviewed-by-paul-merchant; https://web.archive.org/web/20120406042844/http:/sounds.bl.uk/Oral-history/Eminent-scientists/021M-C1379X0013XX-0001V0

[440] Desmond King-Hele, letter to Geoff Perry, 6 April 1966, KET/C/9/29.

[441] Jonathan McDowell, interview, October 2023.

[442] Sarah Mobbs, interview, September 2023.

[443] Ibid.

[444] Ibid.

[445] Ibid.

[446] Martin Mobberley, *It Came From Outer Space Wearing an RAF Blazer!* (Springer-Nature, 2013), 33; Calder Walton, *Spies: The Epic Intelligence War Between East and West* (Simon and Shuster, 2023), 75.

[447] Isabel and Charles Carmichael, interview, September 2024.

[448] Ed O'Grady, interview, November 2024.

[449] Greg Roberts, 'The Amateur and Artificial Satellites,' *Monthly Notes of the Astronomical Society of South Africa*, February 1986: 5.

[450] Ibid., p. 6.

[451] Mark Severance, personal communication, September 2025.

[452] Mark Severance, interview, February 2024.

[453] Mark Severance, interview, October 2025.

[454] 'Listening In,' Video, *60 Minutes*, CBS, 1982.

[455] Gau, 'Sputniks, Bleeps and Mr. Perry.'

[456] Ibid.

[457] Michael Sinnett, interview, September 2023.

[458] Sinnett, 'Members' Tales,' Orbital Focus.

[459] Slater, *Cytringanian Farewell*, 243.

[460] Ibid., 244.

[461] Ibid., 235.

[462] Ibid., 243.

[463] Moore, 'Kettering Connection.'

[464] Geoff Perry, 'The Kettering Group: From Hobby to Profession/Obsession.' Talk at AMSAT UK Annual Conference, 26 July 1997. KET/J/35.

[465] Max White, interview, December 2023.

[466] Moore, Interview for 'The Kettering Connection.' BBC 4 Radio, 1985. KET/J/01.

Bibliography

Books

Astronautics and Aeronautics, 1969: Chronology on Science, Technology, and Policy. Scientific and Technical Information Division, National Aeronautics and Space Administration, 1969.

Beschloss, Michael. *Mayday: Eisenhower, Khrushchev, and the U-2 Affair*. Open Road Media, 2016.

Bulkeley, Rip. *The Sputniks Crisis and Early United States Space Policy: A Critique of the Historiography of Space*. Palgrave Macmillan UK, 1991.

Burgess, Colin, and Hall, Rex. *The First Soviet Cosmonaut Team: Their Lives and Legacies*. Springer New York, 2009.

Burrough, Bryan. *Dragonfly: NASA and the Crisis Aboard Mir*. Fourth Estate, 1999.

Caidin, Martin. *Man Into Space*. Pyramid Books, 1961.

Chertok, Boris. *Rockets and People Vol 2: Creating a Rocket Industry*. National Aeronautics and Space Administration SP-4110, 2000. https://www.nasa.gov/history/history-publications-and-resources/nasa-history-series/rockets-and-people/

Clark, Phillip. *The Soviet Manned Space Program: An Illustrated History of the Men, the Missions, and the Spacecraft*. Orion Books, 1988.

Clarke, Arthur C. *1984: Spring - A Choice of Futures*. HarperCollins Publishers, Limited, 1984.

Phelan, Dominic. *Cold War Space Sleuths: The Untold Secrets of the Soviet Space Program*. Springer, 2013.

Corona Program History. Central Intelligence Agency, 1976. https://www.cia.gov/readingroom/docs/CIA-RDP89B00980R000500070001-2.pdf

Cytringanian Farewell. Old Cytringanians, 1995.

Davidoff, Martin. *The Satellite Experimenter's Handbook.* American Radio Relay League, 1990.

Dickson, Paul. *Sputnik: The Shock of the Century*. Bloomsbury Publishing, 2009.

Ezell, Edward Clinton, and Ezell, Linda Neuman. *The Partnership: A NASA History of the Apollo-Soyuz Test Project.* Dover Publications, 2011.

Ferreiro, Larrie D. *Measure of the Earth: The Enlightenment Expedition That Reshaped Our World.* Basic Books, 2011.

Grahn, Sven. *Dawn, this is Golden Eagle: Listening to Soviet Cosmonauts and Secret Satellites by Radio.* Instant Book AB, 2019.

Hahn, Walter F. *Internal Motives for Soviet Secrecy.* Institute for Defense Analyses, 1963.

Hall, Rex, and David Shayler. *The Rocket Men: Vostok and Voskhod.* Springer Praxis, 2001.

Harford, James. *Korolev: How One Man Masterminded the Soviet Drive to Beat America to the Moon.* Wiley, 1997.

Hutchings, Raymond. *Soviet Secrecy and Non-secrecy.* Barnes and Noble Books, 1987.

International Cooperation and Space Law. U.S. Government Printing Office, 1971.

Ivanovich, Grujica. *Salyut: The First Space Station.* Springer Praxis, 2008.

Jasani, Bhupendra. *Outer Space: Battlefield of the Future?* Taylor & Francis, 1978.

Johnson, Nicholas, and D. M. Rodvold. *Europe and Asia in Space.* USAF Phillips Laboratory, 1991.

Johnson, Nicholas L. The Soviet Year in Space, 1987. Teledyne Brown Engineering, Colorado Springs Office, 1986.

Johnson, Nicholas L. *Soviet Space Programs, 1980-1985.* American Astronomical Society, 1987.

King-Hele, Desmond. *Satellites and Scientific Research.* Routledge and Paul, 1960.

King-Hele, Desmond. *A Tapestry of Orbits.* Cambridge University Press, 2005.

King-Hele, Desmond. *Observing Earth Satellites.* Macmillan, 1966.

Klass, Philip J. *Secret Sentries in Space.* Random House, 1971.

Library of Congress, et al. *Soviet Space Programs: Organization, Plans, Goals, and International Implications: Staff Report.* United States Government Printing Office, 1962.

Library of Congress. Science Policy Research Division. *Soviet Space Programs, 1966-1970: Goals and Purposes, Organization, Resources, Facilities and Hardware, Manned and Unmanned Flight Programs, Bioastronautics, Civil and Military Applications, Projections of Future Plans, Attitudes Toward International Co--peration and Space Law.* U.S. Government Printing Office, 1971. https://babel.hathitrust.org/cgi/pt?id=uc1.a0011467362&view=1up&seq=7

Library of Congress. Science Policy Research Division. *Soviet Space Programs, 1971-75, vol. 1: Overview, Facilities and Hardware, Manned and Unmanned Flight Programs, Bioastronautics, Civil and Military Applications, Projections of Future Plans: Staff Report.* U.S. Government Printing Office, 1976. https://ufdc.ufl.edu/AA00025945/00002

Library of Congress. Science Policy Research Division. *Soviet space programs 1976-1980: Supporting vehicles and launch vehicles, political goals and purposes, international cooperation in space, administration, resource burden, future outlook.* U.S. Government Printing Office, 1982. https://babel.hathitrust.org/cgi/pt?id=uc1.b3907733&view=1up&seq=9

Library of Congress. Science Policy Research Division. *Soviet space programs 1976-1980: Manned space programs and space life sciences.* U.S. Government Printing Office, 1984. https://babel.hathitrust.org/cgi/pt?id=uc1.b4267381&view=1up&seq=9

Library of Congress. Science Policy Research Division. *Soviet space programs 1976-1980: Unmanned space activities*. U.S. Government Printing Office, 1985. https://babel.hathitrust.org/cgi/pt?id=uc1.b4267382&view=1up&seq=9

Library of Congress. Science Policy Research Division. *Soviet Space Programs: 1981-87, vol 2: Space Science, Space Applications, Military Space Programs, Administration, Resource Burden and Master Log of Spaceflights*. U.S. Government Printing Office, 1988.

Lishock, Erika, and Matt Bille. *The First Space Race: Launching the World's First Satellites*. Texas A&M University Press, 2004.

McKay, W. Patrick. *Keep Watching the Skies!* Princeton University Press, 2008.

Miles, Frank, Nicholas Booth, and Geoffrey Perry. *Race to Mars: The ITN Mars Flight Atlas*. Macmillan London, 1988.

Millard, Doug, ed. *Cosmonauts: Birth of the Space Age*. Scala Arts and Heritage Publishers Limited, 2014.

Mobberley, Martin. *It Came From Outer Space Wearing an RAF Blazer! A Fan's Biography of Sir Patrick Moore*. Springer International Publishing, 2013.

Oberg, James E. *Red Star in Orbit*. Random House, 1981.

Ogden, Christopher. *Legacy: A Biography of Moses and Walter Annenberg*. Little, Brown, 1999.

Parkinson, Bob. *Interplanetary: A History of the British Interplanetary Society*. BIS, 2008.

Perry, G. E., and Winterbottom, A. N. *The DRA Table of Space Vehicles*. Defence Research Agency, 1993.

Pocock, Chris. *Dragon Lady: The History of the U-2 Spyplane*. Motorbooks International, 1989.

Prados, John. *The Soviet Estimate: U.S. Intelligence Analysis & Russian Military Strength*. Dial Press, 1982.

Pritchard, David. *The Radar War: Germany's Pioneering Achievement, 1904-45*. P. Stephens, 1989.

Review of the Soviet Space Program, with Comparative United States Data: Report of the Committee on Science and Astronautics. U.S. Government Printing Office, 1967.

Richelson, Jeffrey T. *The Wizards Of Langley: Inside The CIA's Directorate of Science And Technology*. Basic Books, 2008.

Riste, Olav. *The Norwegian Intelligence Service, 1945-1970*. Taylor & Francis, 2014.

Ruffner, Kevin C. (ed.) *Corona: America's First Satellite Program*. U.S. Government Printing Office, 1999.

Sheldon, Charles S. *United States and Soviet Progress in Space: Summary Data*. U.S. Government Printing Office, 1967.

Sheldon, Charles S. "The Soviet space program revisited." In *TRW Space Log 13*, 1974.

Sheldon, Charles S. *A Chronology of Missile and Astronautic Events*. US House of Representatives Report, 1961.

Siddiqi, Asif A. *Challenge to Apollo: The Soviet Union and the Space Race, 1945-1974*. National Aeronautics and Space Administration SP-4408, 2000.

Siddiqi, Asif. *The Death of Vladimir Komarov: Pressure, Politics and Parachutes*. SpaceHistory101.com Press, 2020.

Smith, Arthur. *Planetary Exploration: Thirty Years of Unmanned Space Probes*. Borgo Press, 1988.

Smith, Marcia S. *Space Activities of the United States, Soviet Union, and Other Launching Countries: 1957-1987: Report*. United States, U.S. Government Printing Office, 1988. https://babel.hathitrust.org/cgi/pt?id=pst.000014318771&view=1up&seq=11

Turnill, Reginald, and Arthur Reed. *Farnborough: The Story of RAE*. Hale, 1980.

U.S.-Soviet Cooperation in Space. U.S. Congress, Office of Technology Assessment, OTA-TM-SI-27, 1985.

Sheldon, Charles S. *United States and Soviet Progress In Space: Summary Data Through 1973 and a Forward Look*. Congressional Research Service, Library of Congress, 1974.

Winston, Jay S., et al. *Weather Satellites: Systems, Data, and Environmental Applications*. American Meteorological Society, 1990.

Winterbottom, A. N., and G. E. Perry. *The DRA Table of Space Vehicles 1958-1991*. Defence Research Agency, 1993.

World Armaments and Disarmament: SIPRI Yearbook, 1984. Taylor & Francis, 1985.

World-wide Space Activities: Report. U.S. Government Printing Office, 1977, https://babel.hathitrust.org/cgi/pt?id=uc1.b4265090&seq=7

Video, Audio, Photos

'A Point in Time: The Corona Story,' Video, National Archives and Records Administration, https://archive.org/details/gov.archives.arc.1678526

ATV Today. 'Interview with Geoff Perry, Kettering Grammar School,' 23 April 1971, transcript at KET/H/5/4.

Christy, Robert. 'The tracking station in 1979.' (Much detail on equipment is included in separate links), https://www.orbitalfocus.uk/Kettering/Archive/Equipment.php

Grahn, Sven. 'Kettering Grammar School Satellite Tracking Station,' 1967, Video, YouTube, 1:57, https://youtu.be/UiHh9vOOhmc

Grahn, Sven. Photos with Mark Severance, 1979, and Richard Flagg, 1982, www.svengrahn.pp.se/trackind/Ketterin/Ketterin.htm

Grahn, Sven. Pictures of Geoff Perry, 1945-2000, www.svengrahn.pp.se/trackind/Ketterin/GeoffPix.html

Grahn, Sven. 2016 photos of the Kettering Group exhibit opening at the National Space Centre, www.svengrahn.pp.se/trackind/Ketterin/Leicester2016/Leicester2016.htm

McDowell, Jonathan. Pictures of the 2002 Kettering Group memorial meeting, https://www.planet4589.org/space/kettering/mem02/kett.html

Moore, Patrick. 'Kettering Connection.' Audio, YouTube, 2015. https://www.youtube.com/watch?v=1mk6mSoSjBo

Perry, Geoff and Jos Heyman. Audio, two interviews, 1986. www.svengrahn.pp.se/trackind/Ketterin/Interviews_with_Geoff_Perry

_1986.htm

Perry, Geoff and Reginald Bosanquet. 'ITN Special: Apollo-Soyuz Special.' Video, 4:35, 16 July 1975, https://www.gettyimages.fi/detail/video/apollo-soyuz-special-england-studio-intlive-presenter-uutismateriaali/1289023484?adppopup=true

'Russian space mirror', Stock Footage Video - Getty Images, 1993, https://www.gettyimages.co.uk/detail/video/russian-space-mirror-englandintcms-geoffrey-perry-the-news-footage/868880354

'School Satellite Tracking Station.' Video, 6 minutes, https://www.macearchive.org/films/atv-today-18031966-school-satellite-tracking-station

'Schoolboys Tracking Satellites.' Video, 4:36, from 1966. https://www.youtube.com/watch?v=QWY4nm1bmUU

'Sounds From Space.' http://www.dd1us.de/historical%20sounds%20from%20space.html

'Space Shuttle.' Stock Footage Video Getty Images, 1988, https://www.gettyimages.co.uk/detail/video/london-itn-studio-intcms-geoffrey-perry-interview-sof-the-news-footage/1263565017

'Sputniks, Bleeps, and Mr. Perry:' Video, Youtube.com, 1987.

Sputniks, Bleeps and Mr. Perry part 1
https://youtube.com/watch?v=59s_d7YVC6o

Sputniks, Bleeps and Mr. Perry part 2
https://youtube.com/watch?v=hDmF7bYCZyo

Sputniks, Bleeps and Mr. Perry part 3
https://youtube.com/watch?v=YlG1XYZqjuk

Sputniks, Bleeps and Mr. Perry part 4
https://youtube.com/watch?v=5eQAIrmzsU4

Sputniks, Bleeps and Mr. Perry part 5
https://youtube.com/watch?v=9EdV_oOAovs

'The Schoolboys Who Cracked the Soviet Secret,' Video, PBS, 1989. (Edited version of 'Sputniks, Bleeps and Mr. Perry.') https://archive.org/details/TheSchoolboysWhoCrackedtheSovietSecret

'US Journey Into Space,' Video, Getty Images, 1984,

https://www.gettyimages.co.uk/detail/video/journey-into-space-england-london-intgeoffrey-perry-news-footage/1146703216?adppopup=true

Internet resources

British Film Institute: https://www.bfi.org.uk
British Interplanetary Society: https://www.bis-space.com
British Library: https://www.bl.uk/hubs/research
British Pathé: https://www.britishpathe.com
CIA FOIA Electronic Reading Room: https://www.cia.gov/readingroom/
CIA released documents: http://faqs.org/cia/index.html
Ditton Park Archive: https://www.dittonpark-archive.rl.ac.uk
Geoffrey Perry archive relating to operation of the Kettering Group: https://collection.sciencemuseumgroup.org.uk/documents/aa110093443/geoffrey-perry-archive-relating-to-operation-of-the-kettering-group
Getty Videos archive: https://www.gettyimages.com
GWU National Security Archives: https://nsarchive.gwu.edu
Kettering Group at the National Space Centre: https://www.spacecentre.co.uk/collections/categories/kettering-satellite-tracking-group/
NSA Central Security Service Declassified Documents Declassification & Transparency: https://www.nsa.gov/Helpful-Links/NSA-FOIA/Declassification-Transparency-Initiatives/
Office of Technology Assessment archive: https://www.princeton.edu/~ota/
Old Cytringanians Association: www.cytringanians.org.uk
Orbital Focus: Robert Christy, webmaster. www.orbitalfocus.com
Science and Innovation Park, Science Museum, Wroughton: https://www.scienceinnovationpark.org.uk
Sheldon, Charles II, papers: https://sova.si.edu/record/nasm.xxxx.0141?q=charles+sheldon&t=C
Sheldon, Charles II, articles at globalsecurity.org:

https://duckduckgo.com/?sites=globalsecurity.org&k1=-1&km=l&k9=%230101C4&kj=%23913100&ku=1&kt=h&kr=c&kq=-1&kv=1&q=charles+sheldon&ia=web

Special Collections Research Center, Syracuse University Libraries: https://library.syracuse.edu/special-collections-research-center/

Sven's Space Place. Sven Grahn, webmaster. http://www.svengrahn.pp.se

WorldCat: https://search.worldcat.org

Interviews

Isabel and Charles Carmichael, June 2023, April 2024, September 2024

Mark Severance, February 2024, October 2025

Robert Christy, August 2023

Sven Grahn, August 2023

Richard Flagg, November 2023

Sarah Mobbs, September 2023

Max White, December 2023

Stuart Eves, October 2023

Jonathan McDowell, October 2023

David Dean, April 2024

Michael Sinnett, November 2023

Marcia Smith, October 2023

Martin Davidoff, May 2023

Ed O'Grady, November 2024

Acknowledgements

I had no idea how much help I would need to write this book. While I did virtually all the research and writing, there is much more to publishing a book than that.

The book would not have been written without the incredible efforts of, and support from, Isabel Carmichael, Geoff Perry's daughter, and her husband Charles. They preserved Geoff's papers and artefacts for many years and spent several more years preparing that material for donation to the Science Museum. Isabel and Charles gave me the time and resources that enabled the book to be as detailed and personal as it is. I cannot thank them enough for their most generous assistance.

I would not have been able to work with Isabel and Charles, and all the former Kettering Group members that I interviewed, without the unstinting confidence and encouragement from my great friend of forty years, Mark Severance. He introduced me to Isabel, vouching for me in glowing terms that made me think, 'Wow, I guess I'd better write a really good book!' How Mark knew I was capable of producing this volume, I'll never quite understand—but I will always be grateful for his trust. Mark wrote Appendix B and C, provided technical guidance and wording, and translated Russian language documents.

Two former Kettering Group members had already written about much of the Kettering Group's activities before I came along. Sven Grahn and Bob Christy have, over the years, built a tremendous repository of information about Geoff and the Group on their respective websites: Sven's Space Place (http://www.svengrahn.pp.se) and Orbital Focus (https://www.orbital-focus.uk). Sven and Bob also gave freely of their time and helped me locate materials and people that I likely would not have found otherwise. I thank them for their knowledge, enthusiasm and willingness to share.

Thank you to all the former Kettering Group members and supporters who I interviewed, asked if they would read part or all of the manuscript, and often badgered with numerous follow-up questions: Isabel and Charles Carmichael, Sven Grahn, Robert Christy, Richard Flagg, Mark Severance, David Dean, Sarah Mobbs, Michael Sinnett, Max White, Marcia Smith, Stuart Eves, Jonathan McDowell, Martin Davidoff, Greg Roberts and Ed O'Grady. I also received valuable information and comments via email from Dave Hawkins, Greg Roberts and Derek Slater's sons, William and Michael.

In the course of research, I had wonderful assistance from the following people and organizations: Doug Millard, Doug Stimson, Cate Watson, Jack Garside and Alison Faraday at the Science Museum; Mary Todd at the British Interplanetary Society; Steve Toller at the British Film Institute; Deirdre LaPenna, Amber Mitchell and the Gulf Gate branch staff at the Sarasota County Library; and the staffs at the Kettering Library, the Florida State Library, the British Astronomical Association, the State Library of New South Wales, the National Library of Australia, and the National Library of New Zealand.

I spent countless hours perusing the Kettering Group archives at the Science Museum's National Collections Centre; without the Science Museum's support, this book would not have been possible. The National Space Centre in Leicester has a permanent exhibit of Kettering Group objects that Derek Slater donated and has digitized many of documents that he donated. The Old Cytringanians Association allowed me to use their published material, as did the *Northamptonshire Evening Telegraph,* the British Interplanetary Society and Nigel MacKnight, former editor of *Space Flight News.* Many thanks also to John Percy and Rick Houston for their support and encouragement.

Hugh Barker was my editor, and he suffered through the early drafts of the book with courage while giving me badly needed direction. Karen Hood did the initial design of the book's handsome cover and interior, while Michael Vito Tosto completed the interior layout and the cover. My daughter Kate Welch helped immensely with graphics and illustrations, and my talented son-in-law, Tyler Jacobson, created the gorgeous portrait of Geoff and the Kettering Group on the front cover.

All that is enlightening and entertaining about this book comes from the people and organizations mentioned above. Any mistakes in the book are solely mine. If I have failed to acknowledge someone by name, my apologies, but know that your time and assistance was appreciated.

Eternal thanks and evelasting love to my wife, Vesna Petrovich, for her encouragement, comments and critiques, and for letting me barricade myself in my office for so many years to get this book completed.

About the Author

Desmond Welch is a retired U.S. Air Force fighter pilot and a former NASA Space Shuttle astronaut instructor. He holds degrees in meteorology, physical science and space studies, and is certified in satellite orbit analysis and planning. He lives in southwest Florida with his wife, and enjoys reading, running, and swimming in the Gulf.

You may contact the author at: desmondwelch@europabooks.press

Illustration Credits

Isabel Carmichael: Frontispiece (both), 7, 10, 12, 14, 15, 17, 23, 35, 40, 47, 70, 77, 78, 89, 111, 112, 146, 156, 181, 182, 183, 184 (both), 185, 205

Kate Welch: title page, xxii, 30, 46, 223, 224, 225, 226

Creative Commons (all illustrations from corresponding Wikipedia articles, http://creativecommons.org/licenses/by/4.0): xvi: Sputnik (Andrew Butko, edited), 29: Echo 1, 42: Zenit descent module (Maryanna Nesina), 49: Zenit vehicle, 68: Palmachim (Drink L'Chaim), 99: Jodrell Bank (Mike Peel), 104: Salyut 7, 108: Mir, 115: Prospero (user:geni), 126: Molniya orbit, 129: RORSAT, 154: Znamya

Northampton Evening Telegraph: 24 (top), 32, 36, 64, 71, 73, 186, 219
Robert Christy: 24 (bottom)

British Interplanetary Society: 26, 37, 38, 60, 93, 121, 130, 194, 195, 216

Mark Severance: 65, 102, 108, 127

New Zealand Department of Scientific and Industrial Research: 117

'Geoff Perry' archives at Science Museum: 8, 119 (both)

Sven Grahn: 208

Every effort has been made to properly credit all illustrations used. If any attribution has been omitted or incorrectly listed, please contact Europa Books and it will be corrected promptly.

Index

Page numbers in **bold** indicate an illustration.

A

B

C

D

E

I

J

K

L

M

N

O

P

R

S

T

U

V

W

Y

Z

www.ingramcontent.com/pod-product-compliance
Lightning Source LLC
LaVergne TN
LVHW022027051126
851858LV00004B/63